CONTENTS

Pesticides are poisonous and must be used with caution. Read the label carefully before opening a container. Precautions and directions must be followed exactly. Special protective equipment as indicated must be used.

For information about ordering this publication contact:

Publications
Division of Agriculture and Natural Resources
University of California
6701 San Pablo Avenue
Oakland, California 94608-1239

telephone (510) 642-2431
FAX (510) 643-5470

Bulletin 1911

ISBN 0-931876-34-6
Library of Congress Card No. 88-71015

Printed in the United States of America.

To simplify information, trade names of products have been used. No endorsement of named products is intended, nor is criticism implied of similar products that are not mentioned.

PREFACE

Project W-84 was initiated in 1964 as part of an effort to coordinate segments of work by Agricultural Experiment Stations into regional efforts. Participants in W-84 include entomologists and other scientists from Arizona, California, Colorado, Hawaii, Idaho, Montana, Oregon, Utah, Washington and the US Department of Agriculture.

The purpose of project W-84 is to develop effective biological control procedures and incorporate them into pest management programs. The research work has been concentrated on problems related to biological control through use of natural enemies and mites. This has included a variety of approaches such as techniques for increasing numbers of parasites and predators, modifying agronomic procedures to protect beneficial insects, supplemental rearing and releasing of natural enemies, introducing new biological control agents, and conducting basic studies relating to parasite and predator biology.

This publication is an outgrowth of the research conducted under the project, and the examples used and techniques described are drawn primarily, although not exclusively, from the project. We hope that more interest will be developed in the pest management concept, and that the research findings from W-84 will be put to many uses.

For those interested in research supported by public funds, this publication will also serve as a progress report.

The audience for this book will be varied, but it is intentionally directed toward persons involved with pest management at the applied level. We assume a certain amount of background knowledge related to agricultural pest problems, but not necessarily a high degree of entomological training.

—*The Editorial Committee*

INTRODUCTION

D. W. Davis and J. A. McMurtry

Integrated pest management can be defined as a strategy of pest containment which seeks to maximize natural control forces such as predators and parasites, and to utilize other tactics only as needed and with a minimum of environmental disturbance (Glass, 1975). It is a concept, not a fixed technique. The concept has been developed to the present level over a period of time, and will continue to evolve as long as pest problems exist. Pest management organizes a combination of approaches to pest control into a system that maximizes the advantages while minimizing the disadvantages of each method. Particular emphasis is given to biological control, using beneficial predators and parasites. Each crop will require a different combination of approaches to pest management. Most ideas incorporated in pest management are not new, but there is a new interest in putting the concept to work. This renewed interest in pest management can be attributed to a number of events, including: increased public concern over the extensive use of pesticides, widespread development of insect resistance to pesticides, rise of previously minor pests to major status through the destruction of natural enemies, difficulties in obtaining certain pesticides, and the development of better techniques to evaluate biological relationships.

Chemical control has been attractive in the past because of its ease of application and quick reduction in pest populations. However, pest control cannot be considered only in terms of immediate or short-term values. The effects over a long period of time, or in adjacent areas, may be of more importance than either direct pest control or immediate economic returns.

The pest management approach requires changes in traditional thinking and in certain basic philosophies. Unfortunately, the public has been programmed to think along certain lines. We are faced with slogans such as "the only good bug is a dead bug." At the market place we will not accept produce with superficial injury even though the quality of the edible portion is not affected. The farmer sets his control goals as near to 100 percent as possible, although no economic losses can be shown from low population levels of many pests. If pest management is to work, we must educate both ourselves, and the public in general, to change these attitudes.

There are many examples of pest injury which are strictly superficial. In our effort to get prettier fruit, or greener foliage, we have developed a syndrome of spraying for "cosmetic reasons only." Chemical applications often become excessive, both in cost to the grower and in adverse effect on beneficial organisms. They often do not increase the amount of edible produce. This type of pesticide application must be reduced and the public educated to accept fruits and vegetables with some superficial damage.

Pest management requires a knowledge of the population dynamics of the pest and its natural enemies in relation to cultural practices and plant varieties. All control must be considered within the context of agro-ecosystem and economics of production. Pest management does not exclude the use of pesticides, but attempts to work them into the ecosystem as compatible components of modern agriculture.

Pest management depends on careful monitoring of all significant organisms. Decisions regarding spray applications should not be made by the calendar, but rather on the basis of pest numbers, their developmental stages, their potential for causing damage, and the kinds and numbers of natural enemies. Field samples should be taken at intervals to predict whether biological agents will handle the control adequately. Trained personnel must be available to help with these interpretations.

We must develop a clearer concept of economic injury levels related to pest populations—in other words, how much economic loss can be expected from a particular number of pests at a particular stage of crop development. The decision regarding any control action must be evaluated not only on immediate costs and pest reduction, but on long-range effects as well. An action which results in immediate control may commit a grower to a sequence of procedures which would be very expensive in the long run.

We must learn to accept certain types of noncrop plants around agricultural areas. Some plants should be controlled, but this should not be an automatic rule. Many beneficial insects, including both pollinators and those used in biological control, require habitats or food sources which cannot always be supplied by the crop itself. Selected plant species, especially at field margins, should be encouraged and not sprayed by pesticides.

This book is intended to be a source of pest management ideas. The diversity of experimental approaches is intended to stimulate further exploration. No attempt is made to duplicate material presented in traditional textbooks on pest management or pest control. Many ideas will work on certain crops but not on others, while all of them will have to be modified to fit a wide range of climatic and geographical conditions.

Literature Cited

GLASS, E. H. (Coordinator)
1975. Integrated pest management: Rationale, potential, needs and improvement. Entomol. Soc. Am. Special Publ. 75-2. 141 pp.

Section I
BASIC CONCEPTS OF PEST MANAGEMENT

Chapter 1.

RE-EXAMINATION OF PEST CONTROL PRACTICES

L. A. Andres, E. R. Oatman and R. G. Simpson

Introduction

History of pest control

Early man cultivated the land to increase and assure a supply of food, thereby replacing the mere gathering of nature's produce. Since that time, the balance of plant and animal life in the agricultural communities has become increasingly disturbed and is in a continuous state of change. Agricultural pests have burgeoned, creating new and continuing problems. In the past century, scientific investigations led to remarkably effective methods for the reduction or control of destructive insects, weeds, diseases and nematodes. Paramount in this battle to control pests was the discovery of DDT and 2, 4-D. Control of pests with chemicals has been a major force in the current expansion of the human population in the world. With the development of the synthetic organic pesticides immediately after World War II, the use of pesticides grew at a phenomenal rate.

Factors increasing pest problems

Gradually-increasing labor costs,development of effective but costly farm machinery using more energy, the advent of new pesticides, fertilizer use etc., have altered the economics of food production and forced farmers to acquire more land and concentrate on the production of a single crop rather than on diversified crops on small family farms. This system of monoculture cropping increased pest problems by eliminating the mosaic of habitats which supported many predators and parasites and stabilized pest populations. In addition, breeding for high-yielding, uniform plants has eliminated much plant heterogeneity and resistant insect biotypes, thereby adding to the problem of pest outbreaks. The accidental introduction of foreign insects and weeds, free of their own natural enemies (parasites, predators, and phytophagous insects), has led to pest outbreaks often more serious than those of indigenous pests. For example, recent introductions of the eastern strain of the alfalfa weevil, *Hypera postica* Gyll., and the cereal leaf beetle, *Oulema melanopsis* Linn., have caused tremendous losses to alfalfa and small grains, respectively. Problems to human health and death of livestock have also occurred through the introduction of pests such as the face fly, *Musca autumnalis* De Geer, imported fire ant, *Solenopsis saevissima richteri* Forel, and the weed halogeton, *Halogeton glomeratus* Bieb. All these have contributed to increase in pesticide use. As the ability to control pests increased, so did the demand for higher-quality produce; this has often resulted in a vicious cycle of pesticide use to meet increased public demands.

Problems associated with pesticides

Rather than solving pest problems, the extensive use of pesticides often has resulted in a combination of new problems without eliminating the old ones. A number of individuals foresaw the onset of problems such as pesticide residues, biomagnification (chemical buildup in the food chain), and resurgence created by resistant and secondary pest outbreaks, and attempts were made to alert others to the potential hazards. Nevertheless, pesticides often are still used without regard to detrimental side effects. For example, studies in Colorado show that early treatments with carbofuran (Furadan), widely used to control larvae of the alfalfa weevil, can reduce the emergence of the parasite, *Bathyplectes curculionis* (Thomson), by approximately 60 percent (R. G. Simpson *unpublished data).* In Hawaii, herbicides used to maintain weed-free margins of sugarcane fields led to problems: they eliminate important nectar source plants for the New Guinea sugarcane weevil parasite, *Lixophaga sphenophori* (Villeneuve), (Topham and Beardsley, 1975). We must alter our thinking toward more sound, long-term pest control approaches. This is not to suggest eliminating pesticides, only to reduce the amounts being used and to use them wisely in integrated pest management systems.

Identification of pest problems

The first, and most important, step in pest management is to identify the pest problems. Those involved in crop production should be aware of the key insect, disease and weed pests, and the primary beneficial organisms associated with them. Where the identity is unknown, extension specialists or scientists at state universities often can give the proper identification. Control measures should not be undertaken unless the pest is present in sufficient numbers to cause economic loss (see "Sampling and Monitoring"). Each insect species is distinct in regard to its reproductive potential, dispersal characteristics, number of generations, and damage potential (see "Economic Thresholds and Economic Injury Levels").

Prevention or Reduction of Pest Problems

Once potential insect and weed pests have been identified, the prudent farmer will try to prevent their invasion and buildup in his fields. Just as he might adjust his planting dates to take advantage of favorable weather, his choice of cultural practices, crop varieties, and planting times can minimize his losses from crop pests (see "Cultural Practices in Pest Management").

Planting insect-resistant crops can reduce pest losses. A classic example is the use of Hessian fly *(Mayetiola destructor [Say])* resistant wheat in California. Within several years after the widespread planting of resistant wheat varieties, populations of the fly were so reduced that further studies with this insect and the breeding of resistant strains was abandoned due to a lack of the insect in the area (Painter, 1958). Numerous other resistant crops have been developed and released for use in minimizing insect and pathogen damage. Less-known studies are underway to develop crop plants that will inhibit the germination of weed seeds, thus reducing losses from plant competition.

The farmer's choice of cultural and harvest methods can also influence his losses from insect pests. For example, using herbicides for weed control rather than the traditional French plow in some California vineyards resulted in an increase of the omnivorous leafroller, *Platynota stultana* Walsingham. This was apparently one of the side effects of leaving mummified bunches of grapes, a preferred overwintering site of this insect, on the ground under the vines. Normally, these grapes along with other trash would have been plowed under (Lynn, 1969) (see "Sampling and Monitoring" for details). The practice of harvesting and storing unshelled corn provided an excellent overwintering site for the European cornborer, *Ostrinia nubilalis* (Hübner). The use of the picker-sheller should reduce this annual source of reinfestation since the cobs, in which this insect overwinters, are left in the field and plowed under (Price and Waldbauer, 1975). Many farmers also know that well-timed applications of fertilizer and water can reduce the impact of insect feeding. Green chopping of alfalfa hay gives about 50 percent better alfalfa weevil control than cutting and baling. In a comparison of sprinkler irrigation and flood irrigation, the former substantially reduced alfalfa weevil numbers (Davis, unpublished).

Biological Control

In addition to sound agronomic practices to minimize or prevent the build-up of agricultural pests, farmers should conserve and encourage the naturally occurring enemies of these pests. Traditionally, biological control is defined as the action of parasites, predators, and pathogens in maintaining the density of an organism at a lower level than would occur without these natural enemies. It is a natural phenomenon in which man may or may not play an active role. When he does (applied control), emphasis has been primarily on pest insects and weeds, using insects principally as biological control agents. The subject was recently discussed in considerable detail by van den Bosch and Messenger (1973) and DeBach (1974).

Classical biological control is the control of an exotic pest species by introduced natural enemies. This involves foreign exploration, quarantine processing of collected material, mass propagation of the natural enemies, their field colonization, and evaluation of their impact on the pest population. Nearly 100 species of pest insects and weeds have been completely or substantially controlled by imported natural enemies (van den Bosch and Messenger, 1973).

One of the first and best examples of biological control in the U.S. was the introduction of a coccinellid, *Rodolia cardinalis* (Mulsant), and a tachinid, *Cryptochaetum iceryae* (Williston), from Australia in 1888 to control the cottony-cushion scale, *Icerya purchasi* (Maskell), on citrus in southern California. A recent outstanding example is the biological control of the walnut aphid, *Chromaphis juglandicola* (Kaltenbach), on English or Persian walnuts in central and northern California in 1970 by an Iranian strain of the aphidiine wasp, *Trioxys pallidus* (Haliday) (van den Bosch and Messenger, 1973).

Two of the best known and early examples of the biological control of weeds were the control of prickly pear cacti, *Opuntia* spp., in Australia by a padboring caterpillar aided by a mealybug, and control of the Klamath weed, *Hypericum perforatum* Linn., in northern California by leaffeeding beetles (National Academy Science, 1968). More recent examples of classical biological control of exotic weed pests were discussed by Andres and Goeden (1971).

The above examples of biological control involved the importation of natural enemies from the native home of the pest. However, certain species of native parasites and predators may also be preserved or increased through such practices as using supplemental foods or attractants.

Pathogens of insect pests have been used most successfully for biological control when developed as microbial insecticides (National Academy Science, 1969*a;* Heimpel, 1974). The first such pathogen was the milky disease, *Bacillus popilliae* Dutky, used for control of the Japanese beetle, *Popillia japonica* Newman, in the eastern U.S. The most widely used commercial microbial insecticide consists of various strains of *Bacillus thuringiensis* Berliner which are effective against many lepidopterous pests. More recently, two nuclear polyhedrosis viruses have been approved for use and others produced commercially for experimental use against lepidopterous pests such as the corn earworm, *Heliothis zea* (Boddie), cabbage looper, *Trichoplusia ni* (Hübner), the beet armyworm, *Spodoptera exigua* (Hübner), and the Douglas fir tussock moth, *Orgyia pseudotsugata* McDonnough.

The principal advantage of biological control is permanent suppression of pest species, which makes it one of the most effective methods of control. Its self-perpetuating characteristics also make it one of the least costly methods. Moreover, its contribution to environmental quality is inestimable and invaluable.

The future of biological control depends to a large extent on the availability of financial support. Success in classical biological control is directly proportional to the amount of effort given to natural enemy introductions (DeBach, 1964). Detailed studies of biological relationships are necessary to increase the parasites and predators. Thus, man's increased ecological concern must be matched by increased funds if biological control is to reach its fullest potential.

Physical and Mechanical Control

Oftentimes a pest may be controlled mechanically or physically by removing, excluding it from the crop, or in some way interrupting its growth or reproduction. When only a few weeds or pest animals are involved, the farmer may remove them by hand. Where the pest has not yet gained entry to a field, exclusion devices may be employed. For example, screens have been recommended to remove weed seeds from irrigation canals (Kelley and Bruns, 1975). Aluminum foil barriers have been used to exclude salt marsh caterpillars *(Estigmene acrea* [Drury]) from lettuce fields in southern California. Reflective foils repel some insects from crop areas. Johnson *et al.* (1967) reported a reduction in the number of aphids on gladiolus through the use of such foils and white polyethylene mulch.

Trapping by various means may aid in insect control. Steiner *et al.* (1965) reported the eradication of the Oriental fruit fly, *Dacus dorsalis* Hendel, from a Pacific island by luring the male flies with methyl eugenol to poison-impregnated fiberboard blocks. In this way, they prevented fertilization of the eggs carried by the females, which led to the elimination of the species.

Mulching with straw, sawdust, wood chips and other materials, including black polyethylene plastic also has proven useful in retarding weed infestations.

Control by physical and mechanical means must often be tailored to the specific problem at hand. Their chief advantage is that they do not contaminate the environment.

Chemical Control

Pesticides appear in so many forms and under a multitude of names, and are recommended for the control of so many pests, that many people are confused. These individuals must rely on the dealer or farm supplier for advice which may or may not be accurate. A little study or review can be invaluable for the extension worker or grower to prevent mistakes or even serious accidents. The requirement for certification before applying many of the pesticides should help overcome mistakes related to uses.

Pesticide formulations

A pesticide is composed of a toxicant which is listed on the container as the active ingredient, and one or more inert substances that provide specific functions in the formulation. The inert substances may act as carriers, diluents, emulsifiers, spreader-stickers, dispersants, etc. Generally insecticides are formulated as: (1) **dusts,** which are used dry and mixed with or impregnated on organic substances such as walnut shell flour or pulverized minerals such as bentonite, talc, prophyllite, and attapulgite; (2) **granules,** which are similar to dusts except for particle size and are used as soil dressing (on top or below the surface). Granules are usually applied with special applicators or with fertilizer spreaders; (3) **wettable powders,** which appear like dusts but are to be diluted and suspended in water then used as a spray; (4) **emulsifiable concentrates,** the most common formulation, composed of the toxicant, a solvent, and an emulsifier. When the concentrate is mixed with water, an emulsion is formed of the oil-in-water type, which can be used as a spray; (5) **fumigants,** which are insecticides used in the gaseous form in areas where they can be confined such as storage bins, holds of ships, buildings, and greenhouses. These are formulated as liquids under pressure which change into a gas when released from their containers.

Classes of pesticides

Classification of the pesticides usually is based on the chemical nature of the material. The major divisions are inorganic and organic (Fronk, 1971). The inorganic pesticides include sulfur, lime sulfur, fluorides, coppers, and arsenicals but they have been replaced, largely, by newer more effective synthetic organic materials. Organic pesticides can be subdivided into oils, botanicals, and synthetics. Oils are obtained from petroleum and used primarily against pests on tree fruits as herbicides, and as carriers for other pesticides. Botanicals are derived from plants and include materials such as nicotine, cube, rotenone, pyrethrum, sabadilla, and ryania. The synthetic organic pesticides include the more important materials in use today. Insecticides often are classed as chlorinated hydrocarbons, organophosphorus and carbamate compounds. The properties of these groups have been discussed by many authors including Metcalf (1966) and Mitchell (1966).

Hazards

Pesticides are toxic materials and should be treated as such. Most problems with their use are the result of human error or poor judgment. There are several points of particular concern: (1) careless mixing and filling the sprayer or duster, (2) excessive drift away from the target areas during the application, (3) residues after application in the field, and (4) improper storage of materials. The precautions that must be followed to avoid accidents can be obtained by reading the label on the container. Necessary warnings are required by law to be on the label.

Toxicities

Pesticides vary in their toxicities to mammals and other nontarget organisms. These are determined by experiments on mice, rats, rabbits, fish, birds, and occasionally dogs. Toxicities are determined on the acute oral toxicity, acute dermal toxicity, inhalation toxicity and chronic feeding toxicity. Acute oral and dermal toxicities are usually expressed in terms of LD_{50}, the theoretical dose required to kill 50 percent of the test animals. LD_{50} is expressed in milligrams of toxicant per kilogram of body weight (mg/kg) of the test animal. A lower figure indicates greater toxicity. For example the LD_{50} of methoxychlor is 6000 while that of dieldrin is 100. This shows that, under identical testing conditions, dieldrin is 60 times more toxic than methoxychlor. Inhalation toxicities are expressed as LC_{50} (ug/1) and chronic toxicities as parts per million (ppm) of the daily diet for a specified number of days. In addition to toxicity, several other factors such as teratogenicity, oncogenicity, and genetic effects, must also be determined. The relative toxicity and hazard levels of an insecticide are indicated on the labels as "Poison", "Warning", or "Caution".

Poison-control centers

All states maintain one or more Poison Control Centers where emergency treatment or information can be obtained. All persons involved with pesticides should keep the telephone number of the PCC readily available.

Manipulation of chemical treatments

Pesticides can control many pest outbreaks and save billions of dollars, but they can have adverse ecological effects. In recent years, there has been a shift from the more persistent materials to those that are less persistent but often more acutely toxic, This trend should reduce certain problems, but with several pesticides we will have to be more concerned with acute toxicity to man and other organisms.

There are a number of ways that chemical treatments can be altered or manipulated to avoid large losses of nontarget species of insects (see "Selective Pesticides"). Some of the more important methods are: (1) pesticides should be applied based on field samples to insure that damaging numbers of pests are present rather than on a predetermined schedule (see "Economic Thresholds and Economic Injury Levels"); (2) systemic or granular insecticides should be used when possible; (3) applications should be made in accordance with the insecticide tolerant stages of the beneficial insects. Generally, adult parasites and predators are subject to higher direct mortality from pesticides than are immatures (Bartlett, 1964). Croft and Brown (1975) reported that parasites are most susceptible to insecticides than are their hosts or predators; (4) insects seldom infest fields or crops uniformly, so pesticides should be applied only to portions of fields, in spots or strips when possible; (5) pesticide applications may be timed to coincide with periods when beneficial insects are not present on the crop or when they are in protected sites; (6) selective insecticides that favor natural enemies, yet control the pest, should be used if available; and (7) dosages of pesticides should be reduced to favor natural enemies of the pest insect.

Other Control Methods

Besides natural enemies, other biological methods have been used successfully to control pest insects. The more common ones include sterility, pheromones, hormones, and the development of resistant varieties.

Sterility

The sterile-insect (autocidal) control technique involves rearing large numbers of the pest species, sterilizing them with gamma radiation or chemosterilants, and then releasing them in overflooding numbers into a natural (untreated) population, where the treated males mate with normal females, yielding nonviable eggs. This results in a lower progeny production for the next generation. By repeating the releases for consecutive generations, the probability for fertile matings becomes less and less and eventually the natural population is controlled (National Academy of Sciences, 1969*b*). The technique was used to eradicate the screwworm fly, *Cochliomyia hominivorax* (Coquerel), on the island of Curacao and in Florida. It is annually used in southwestern U.S. and northern Mexico to manage this pest (Knipling, 1972). Sterile Mexican fruit flies, *Anastrepha ludens* (Loew), and pink bollworms, *Pectinophora gossypiella* (Saunders), adults are released annually along the Mexican-California border and in the southern San Joaquin Valley of California, respectively, to prevent the establishment and spread of these pests in these areas. Feasibility studies are now being conducted into two major fruit insects in North America, the codling moth, *Laspeyresia pomonella* (L.), and the western cherry fruit fly, *Rhagoletis indifference* (Curran).

The sterile-insect release method is not feasible for all pests. It works best when the pest population occurs at a low level and with those insects that mate only once. The method is therefore most effective when used in integrated control programs in which the population has first been reduced by other methods (National Academy of Science, 1969*b*). It has been used to eradicate newly introduced pests, such as the Mediterranean fruit fly in California before it became widely distributed.

Chemosterilants also may be used in baits to induce sterility in pest insects which are not amenable to mass rearing and release. To control or eradicate such species, it is necessary to induce sterility in a large proportion of an existing natural population in such a way that the sterile males will mate with (and thus render infertile) the females which escape the sterilizing chemical. Pest species which are attracted to baits in large number are especially susceptible to this method of control.

The principle disadvantage of chemosterilants is that many of those which are most promising for practical insect pest control are hazardous to man and higher animals, being mutagenic, carcinogenic, or teratogenic agents. Thus, it is imperative that sterilants be used in such a way that they will not be contacted by man or beneficial animals (National Academy of Sciences, 1969*b*). When used for sterilization of native populations in the field, they should not be used on a large scale until less hazardous chemicals are produced or until safer techniques are available (Proverbs, 1969).

Pheromones

Pheromones are chemicals used for communication between two or more animals of a single species (Shorey, 1973). As a class of biologically active compounds, attractant pheromones have several characteristics which make them potentially useful as nontoxic pest control agents. These are: activity in minute quantities; relative specificity; low toxicity; minimal adverse effects on non-target organisms, such as beneficial insects; and potentially low cost of development (Tette, 1974).

Pheromones have been used successfully for the detection of pest insects such as the codling moth, *Laspeyresia pomonella* (L.), boll weevil, *Anthonomus grandis* Boheman, and the gypsy moth, *Porthetria dispar* (L.). Both the natural source of attractant pheromone from virgin females and synthetic pheromones are being used effectively to assess the threat of a pest, to pinpoint incipient infestations, and to provide more precise timing of pesticides. Thus, they are useful in integrated control programs, especially for fruit orchards. Pheromone traps also have been used successfully for detecting the spread of pests such as the California red scale, *Aonidiella aurantii* (Maskell), and the gypsy moth into new areas where, if detected early, eradication measures may be taken (Tette, 1974).

The use of pheromones to control pest insect populations is based on either disruption of the insect's communication system so that mating is not accomplished, or on the removal of a large proportion of the adult population by mass trapping.

The first method (pheromone disruption) involves continuously saturating the environment in a given area with pheromone from multiple emission sources so that male insects are unable to orient to and locate the females. This has been tested against several insects such as the pink bollworm, cabbage looper, and the gypsy moth with varying degrees of success. The mass-trapping method uses pheromones to attract the insect to a trap where it is killed either by a physical restraint, such as a sticky surface, or by an insecticide combined with the pheromone. The method has been used for orchard pests, the boll weevil, and for bark beetles with varying degrees of success (Tette, 1974).

Programs utilizing pheromones for survey or control are discussed at length by Birch *et al.* (1974).

Hormones

Molting is a characteristic feature of insect growth and development. Three types of hormones control molting and metamorphosis in insects: the brain hormone, molting hormones (or ecdysones), and the juvenile hormones. Hormones are also involved in the regulation of insect diapause (Robbins, 1972).

Desirable characteristics of hormonal insecticides include high biological activity, specificity, and low vertebrate toxicity. Juvenile hormone compounds have the greatest immediate potential for insect control. They readily penetrate the insect cuticle and some act as fumigants. Thus, their activity is similar to conventional insecticides. The principal lethal effect of compounds with juvenile hormone activity is to interfere with the larval-pupal, pupal-adult, and nymphal-adult molts. The untimely presence of one of these compounds during these molts produces abnormal forms which usually fail to undergo normal ecdysis, are incapable of normal reproduction, and soon die (Bowers, 1971).

The principal disadvantage of juvenile hormone compounds for insect control is that they do not block the development of or kill the growing larvae or nymphs which usually are the most destructive stages. Thus, to be effective in the field, repeated applications of juvenile hormone mimics will have to be made or highly persistent compounds will have to be used (Robbins, 1972).

Though several small scale field plot and cage tests have been performed with hormonal insecticides, most studies have been conducted under laboratory conditions. Thus, their effectiveness under field conditions is still largely unknown.

Resistant varieties

Plant susceptibility to insect attacks varies from variety to variety and individual to individual of the same species. This is due to the relative amount of heritable qualities possessed by the plant which influence the ultimate degree of damage caused by the insect (Painter, 1958). These qualities enable some plants to thrive and reproduce under severe insect attacks. The degree of resistance may vary from a complete immunity to high susceptibility.

Painter (1958) classified the plant resistance to insects into 3 categories based on the mechanisms involved. These are, nonpreference (for oviposition, food, or shelter), antibiosis (adverse effect of plant on biology of insects), and tolerance (repair, recovery, or ability to withstand infestation). Some resistant varieties might utilize two or more components.

From the early discovery of resistance in an apple variety (Winter Majetin) to the woolly aphid, *Eriosoma lanigerum* (Hausin) to the present day, utilization of resistant plants has been a major control against many insects and diseases. Because of low costs, cumulative reduction in pest damage, reduced environmental contamination and energy use, and the ease with which this method can be applied under field conditions, the resistant varieties indeed offer one of the excellent methods of pest control. However, it may not be available or feasible for all insect pests and, considering the amount of time involved in breeding, it may be uneconomical for some insects. Nevertheless, for many crops this method can form the basis for a sound integrated pest management.

Integrated Control and Pest Management

The trend for the immediate future is clearly in the direction of integrated control (integrated pest management). The terms have often been used interchangeably. Originally, integrated control meant the blending of biological and chemical pest control to enhance their mutual effectiveness (Luckmann and

Metcalf, 1975). For example, when the spotted alfalfa aphid, *Therioaphis maculata* (Buckton), became a problem the number and variety of natural enemies were assessed, then considered along with the age and condition of the crop and its market value before applying carefully selected pesticides. Every effort was made to reduce crop losses by enhancing the action of the natural enemies against existing and potential alfalfa pests.

The term pest management, as originally used, was broader than integrated control in that it included the selection and manipulation of all pest control tactics in a manner that best met the needs of both the producer and the consumers. It included planting schedules, spatial arrangement of crops, selection of pest resistant varieties, use of natural enemies, use of physical, mechanical, or chemical controls, and various other environmental, ecological, and sociological considerations. It also requires extensive knowledge and analysis of the agroecosystem. Philosophically, there is little need to distinguish between the terms integrated control and pest management. Ideally, the farmer should integrate all control tactics at his disposal, using nonchemical techniques wherever possible. Hopefully, he will adopt pest management concepts to prevent most problems from developing, rather than trying to control them after they get out of hand.

Literature Cited

ANDRES, L. A., and R. D. GOEDEN
1971. The biological control of weeds by introduced natural enemies, pp. 143-164. *In:* Biological control. C. B. Huffaker, ed. Plenum Press (New York-London). 511 pp.

BARTLETT, B. R.
1964. Integration of chemical and biological control. *In:* Biological control of insect pests and weeds. Paul DeBach ed. London: Chapman and Hall, 844 pp.

BIRCH, M., K. TRAMMEL, H. H. SHOREY, L. K. GASTON, D. D. HARDEE, E. A. CAMERON, C. J. SANDERS, W. D. BEDARD, D. L. WOOD, W. E. BURKHOLDER, and D. MULLER-SCHWARZE.
1974. Programs utilizing pheromones in survey or control, pp. 411-461. *In:* Pheromones. M. C. Birch, ed. North-Holland Research Monographs: Frontiers of Biology, Vol. 32. North-Holland Publ. Co. (Amsterdam-London). 495 pp.

BOWERS, W. S.
1971. Juvenile hormones. pp. 307-30. *In:* Naturally occurring insecticides. M. Jacobson and D. G. Crosby (eds.). M. Dekker, Inc., New York. 585 pp.

CROFT, B. A., and A. W. A. BROWN.
1975. Responses of arthropod natural enemies to insecticides. Ann. Rev. Entomol. 20:285-335.

DEBACH, P., ed.
1964. Biological control of insect pests and weeds. London: Chapman and Hall, 844 pp.

DEBACH, P.
1974. Biological control by natural enemies Cambridge University Press. New York. 323 pp.

FRONK, W. D.
1971. Chemical control, pp. 191-217. *In:* Fundamentals of applied entomology. R. E. Pfadt, ed. The MacMillan Company. New York. 693 pp.

HEIMPEL, A. M.
1974. The development and progress of insect microbial control, pp. 528-31. *In:* Proceedings of the summer institute of biological control of plant insects and diseases. Maxwell, F. G. and F. A. Harris (eds.). Univ. Press of Miss., Jackson. 647 pp.

JOHNSON, G. V., A. BING and F. F. SMITH.
1967. Reflective surfaces used to repel dispersing aphids and reduce spread of aphid-borne cucumber mosaic virus in gladiolus. J. Econ. Entomol. 60:16-18.

KELLEY, A. D. and V. F. BRUNS.
1975. Dissemination of weed seeds by irrigation water. Weed Science 23(6):486-93.

KNIPLING, E. F.
1972. Sterilization and other genetic techniques, pp. 272-87. *In:* Pest control strategies for the future. National Academy Sciences. Washington, D.C. 376 pp.

LUCKMANN, W. H., and R. L. METCALF.
1975. The Pest-Management Concept. pp. 3-35. *In:* (R. L. Metcalf and W. H. Luckmann, eds.) *Introduction to Pest Management,* John Wiley and Sons, New York, 587 pp.
LYNN, C. D.
1969. Omnivorous leaf roller, an important new grape pest in the San Joaquin Valley. Calif. Agric. 23(4):16-17.
METCALF, R. L.
1966. Requirements for insecticides of the future, pp. 9-23. *In:* Pest control by chemical, biological, genetic, and physical means, A symposium. Published as U.S. Dept. Agr. ARS-33-110, July.
MITCHELL, L. E.
1966. Pesticides: properties and prognosis, pp. 1-22. *In:* Advances in Chemistry Series. Vol. 60. R. F. Gould, ed. American Chemical Society. 309 pp.
NATIONAL ACADEMY SCIENCE.
1968. The biological control of weeds, pp. 86-119. *In:* Principles of plant and animal pest control. Vol. 2. Nat. Acad. Sci. Publ. 1597. 471 pp.
1969a. Microbial control of insects, pp. 165-95. *In:* Insect-pest management and control. Principles of plant and animal pest control. Vol. 3. Nat. Acad. Sci. Publ. 1965. 508 pp.
1969b. Sterilization, pp. 324-59. *In:* Insect-pest management and control. Principles of plant and animal pest control. Vol. 3. Nat. Acad. Sci. publ. 1695. 508 pp.
PAINTER, R. D.
1958. Resistance of plants to insects. Ann. Rev. Entomol. 3:267-90.
PRICE, P. W. and G. P. WALDBAUER.
1975. Ecological aspects of pest management, pp. 37-73. *In:* (R. L. Metcalf and W. H. Luckmann, eds.). "Introduction to Pest Management", John Wiley and Sons, New York. 587 pp.
PROVERBS, M. D.
1969. Induced sterilization and control of insects. Ann. Rev. Entomol. 14:81-102.
ROBBINS, W. E.
1972. Hormonal chemicals for invertebrate pest control, pp. 172-96. *In:* Pest control strategies for the future. National Academy Sciences. Washington, D.C. 376 pp.
SHOREY, H. H.
1973. Behavioral responses to insect pheromones. Ann. Rev. Entomol. 18:349-80.
STEINER, L. F., W. C. MITCHELL, E. J. HARRIS, T. T. KOZUMA and M.S. FUJIMOTO.
1965. Oriental fruit fly eradication by male annihilation. J. Econ. Entomol. 58:961-64.
TETTE, J. P.
1974. Pheromones in insect population management, pp. 399-410. *In:* Pheromones, M. C. Birch, ed. North-Holland Research Monographs: Frontiers of Biology, Vol. 32. North-Holland Publ. Co. (Amsterdam-London). 495 pp.
TOPHAM, M., and J. W. BEARDSLEY.
1975. Influence of nectar source plants on the New Guinea sugarcane weevil parasite, *Lixophaga sphenophori* (Villeneuve). Proc. Hawaii. Entomol. Soc. 22:145-54.
van den BOSCH, R. and P. S. MESSENGER.
1973. Biological control. In-text Educ. Publ. New York and London. 180 pp.

Chapter 2.
SAMPLING AND MONITORING

J. W. Beardsley, M. T. AliNiazee, and T. F. Watson

Introduction

Sampling of arthropod populations within, or adjacent to, agricultural operations is a prerequisite to understanding population trends and significance of both pest and beneficial species. Intelligent decisions concerning pest control strategies are directly dependent upon knowing the status of all important species and how pest population levels relate to economic thresholds.

In sampling we attempt, by inspecting or counting only small parts of much larger populations of insects and other arthropods on crop plants, to estimate absolute levels of arthropod populations and/or pest damage. Monitoring is simply sampling or making observations periodically to determine population trends. Ultimately, the population or damage estimates derived from these samples or observations are used as a basis for pest management decisions.

Sampling and population monitoring require diverse techniques depending on the crop and the characteristics of the pests and beneficial species involved. Regardless of the procedures required to adequately sample populations, two requisites are essential: practicality and reliability. By practicality we mean methods that can be utilized by anyone involved with a particular crop without undue expenditures for expensive equipment, and without being overly burdensome from the standpoint of time required. Reliability simply means that the method or methods employed will yield reliable numbers that can be used as indices relative to absolute populations and to the damage anticipated from such populations. Samples, therefore, must be "adequate" (i.e.: composed of enough units to permit some type of statistical analysis which will provide the answers we seek, within an acceptable margin of error).

During the past few years various reviews have appeared which deal with the subject of population sampling (Morris, 1960; Southwood, 1966; Gonzalez, 1971; Ruesink and Kogan, 1975). Some of the published information is somewhat theoretical, but more recent publications stress the applied aspects and the importance and utilization of sampling methods in pest management programs. The purpose of this paper is to look at the practical aspects of population sampling and monitoring as related to the decision-making process in pest management. Therefore, the statistical and analytical aspects of sampling will not be covered. These can be found in such references as Cochran (1953), Finney (1953), or Hansen *et al.* (1953).

The mechanics of sampling depend upon the life history and habits of the species involved and on the type of crop. Morris (1960) stressed three main aspects of sampling design: selection of the universe; selection of the sample unit; and determination of the optimum stratification and distribution of sample units within the selected universe. After considering these basic criteria in designing the sampling system for the species and crops involved, the mechanics of collecting, extracting, and counting the arthropods must be worked out in detail. For example, collecting the sample may be either direct or indirect, according to Morris (1960). The direct method involves collecting the target arthropod itself, while the indirect method means collecting a sample of the insect's medium, such as foliage, for later extraction.

The second step in the assessment of a sample is extraction of the insects or other arthropods from the sample unit. For example, the removal of spider mites from a sample of rose leaves with a mite-brushing machine would involve a mechanical method. Other methods include: hand-picking, a chemical method to anesthetize or repel the insects, or a behavioral response to some physical stimulus such as light or heat.

The last step in sample assessment involves counting or measuring (weight or volume) arthropods to provide quantitative data for correlation with pre-established injury levels or for determination of population trends. This is the end result of sampling and provides the data needed on which decisions are based.

In certain situations it may not be necessary or even desirable to sample for the arthropod itself. Sampling may be directed toward the amount of damage or other signs of activity such as the amount of frass present. The steps discussed above, or some modification of them, would still pertain.

Factors Affecting Sampling Error

An important consideration in designing any sampling regimen is the avoidance of bias, either conscious or unconscious, in the selection and collection of sample units. This may be particularly important in large-scale programs involving data obtained by different individuals. In such programs the standardization of sampling procedures by means of carefully prepared detailed instructions is essential. The validity of pest management decisions ultimately depends on the quality and uniformity of the sampling procedures employed. Since no two individuals ever sample in exactly the same way, the simpler and more carefully standardized the sampling methods the smaller will be the sampling error due to differences between individuals. In pest management programs, we need to stress the importance of utilizing personnel who are adequately trained and thoroughly familiar with the sampling procedures to be employed.

Ambient environmental conditions are important sources of sampling error or bias which need to be carefully considered in any program. For example, time of day during which samples are collected can be of major importance. Dumas *et al.* (1962), in a study on the effect of time of day in surveys of predaceous insects in field crops, found that counts taken during early morning or in the evening differed significantly from those taken at mid-day. In sampling populations of the ant *Pheidole megacephala* (Fabr.), in Hawaiian pineapple fields, McEwen and Beardsley (unpublished) found that foraging activity was much greater during the late afternoon and evening than during the middle of the day, and counts of ants attracted to a honey bait were significantly greater when made after 6:00 p.m. Such data suggest that, to minimize sampling error, many kinds of sample units should be collected during the same period on each sampling day. The time of day may be of no importance in sampling populations of sessile insects such as scales, mealybugs, and the like. Many other environmental factors, such as wind, precipitation, humidity, cloud cover, soil temperature, cultural practices and conditions in surrounding habitats or adjacent fields, may affect arthropod activity, and therefore may cause sample bias. This is sometimes unavoidable.

A further important factor which should be considered in any sampling or monitoring program is the spatial distribution of the organisms being sampled. Frequently, the distribution of pest species is more or less at random with very low densities, but tends to become clumped or aggregated as population levels increase (Kuehl and Fye, 1972). This may be true not only of colonial insects such as aphids, mealybugs and the like, but also of many more mobile species. Gonzalez (1971) reported this type of change in distribution with the cotton bollworm *(Heliothis zea* [Boddie]) in California. Obviously, when dealing with pests which tend to develop foci of infestation (clumped distribution) which may rapidly increase to damaging levels, the sampling or monitoring scheme should be sufficiently intense to detect the presence of such foci at an early stage. Spatial distribution also is an important consideration in determining what part or stratum of the crop to sample.

Before any large-scale sampling program is initiated it is well to have a competent statistician review the proposed sampling procedures, so that you may be reasonably certain the data which you will obtain will be adequate to provide the answers needed to make confident pest management decisions. This may also save effort, as many workers tend to take more or larger samples than necessary.

Sampling methods, and the timing of the sampling events themselves, are obviously critical considerations in any pest management system, and may be sources of serious sampling errors. Proper selection of the method used and the sampling intervals are contingent upon a thorough knowledge of the biology and ecology of the target arthropods and the crop to be sampled.

Sampling Methods

Sampling methods may be designed to assess population levels of pest and/or beneficial species or to assess damage caused by the pest. Regardless of whether the pest or its damage is assessed, yield samples also are commonly collected to complement data concerning pest levels or damage. The yield data can then be correlated with pest levels occurring at an earlier date.

Ruesink and Kogan (1975) divided sampling methods used in insect pest management into three primary types: absolute methods, relative methods and population indices. Absolute methods yield estimates of population density per unit of land area in the habitat. Such methods are the best for obtaining the kinds of data needed in detailed population ecology and population dynamics studies. Presumably, all or virtually all individuals of the target species in the sample unit area are collected with these methods. Examples of absolute methods sometimes used in pest management programs are soil samples (such as are collected by means of a soil core sampler), sod samples (e.g., 1 square foot), litter samples, 1 square foot (or other unit of area) of a grain or forage crop, and whole plant samples of row crops (where the number of plants per unit area is known). Generally such methods are relatively non-selective, are often laborious to collect, and usually require considerable processing to extract the desired arthropods.

Relative sampling methods yield density estimates based on units other than land area, which cannot easily be converted to land area. Sampling methods such as the sweep net, visual searching, and most kinds of insect traps are relative methods. Such methods may be quite selective, and generally less time and effort is required to collect and process the samples than with most absolute methods. Relative sampling methods are widely used in pest management programs.

With population indices, the target arthropods themselves are not counted, but products of the target species (such as frass volume or plant damage) are measured. Damage indices of various sorts are commonly used in pest management work.

When relative samples or population indices are employed, sampling errors can be introduced due to differences between individual samplers and differences in the observability or collectability of the target arthropods which result from changes in behavior, time of day, weather, or the height or density of the crop.

It sometimes becomes necessary to convert data obtained by relative sampling methods into absolute units to develop economic threshold levels, etc. A point-sample is one which is designed to relate the number of arthropods or arthropod damage to the number of plants, or plant parts per unit area. For example, Lincoln *et al.* (1963) described a point-sample method of scouting for cotton boll weevils. The scout is instructed to examine the first fifty squares, one-fourth inch in diameter or larger, at each sample site, and to measure the number of row-feet required to provide this sample. The number of squares punctured or damaged by boll weevils is recorded. This procedure is carried out in a least four locations in each average-sized field being sampled. From these data it is possible to estimate the total squares of this size class, as well as damaged squares, on a per acre basis.

Various kinds of sampling methods and sampling devices used in pest management programs are discussed in more detail later in this chapter.

The Sample Unit

Selection of an appropriate sample unit (i.e., that which is counted) is of major importance to the conduct of any pest management program. Ideally, sample units should provide the data needed for decision making, with a minimum amount of extraneous material (debris, arthropods unrelated to the problem under study, etc.). A thorough knowledge of how the target species are distributed on or within the crop plant facilitates the development of stratified sampling techniques, whereby sample units are collected from a given stratum (particular part, level or region of the host plant where the target species is most concentrated). When this is done, only the variation within the stratum sampled contributes to sampling error. Sometimes the strata containing the target species may change with time (i.e., with the age of the crop or stage of development of the target arthropods), and sampling procedures may have to be adjusted accordingly (Southwood, 1966).

In general, the types of sample units commonly used in pest management work can be classified into four main categories.

- Units most directly related to plot area, or volume (e.g., one sweep of a sweep net, 1 square foot of sod, a 10-foot diameter circle of sugarcane, 1 cubic foot of soil, a square-foot D-vac sample of alfalfa). Most such units, except for the D-vac and sweep net, are obtained by use of absolute sampling methods. The D-vac and sweep net methods normally collect something less than 100 percent of the target species, depending upon crop height, density, etc.

- Units most directly related to the individual host or crop plant or plant part as arthropod number per host unit, or arthropod damage per host unit. For example, we may count the number of insects per leaf, per leaf section, per shoot, per fruit, per branch, per section of trunk, per stalk, per unit weight of leaves, twigs, roots, fruits, etc. Most such sample units are obtained by relative methods, except in the case of whole plant samples where the number of plants per unit area can be determined. Or, we may measure damage in terms of percent of leaf surface eaten, percentage of bored internodes in sugarcane, volume of insect frass, feeding spots per fruit, etc. These latter samples are used to obtain damage indices.

- Units most directly related to time (e.g., a 3-minute D-vac sample, a 24-hour light trap or pheromone trap sample, a 1-week pitfall trap or sticky-board trap sample, etc.). Methods used to obtain such samples generally are relative methods.

- Units most directly related to the pests themselves, such as are often used in evaluating natural enemy populations (e.g., parasites per hundred eggs, or per egg mass, percentage of parasitized larvae or pupae, predator/prey ratios, etc.). To be truly meaningful, such units also need to have the host density quantified in some manner.

In any sampling program the sample unit chosen will, of course, depend upon the information being sought. Thus, if we wish to analyze the entire fauna associated with a particular crop plant, we will need to sample entire plants, using an absolute method such as that proposed by Leigh *et al.* (1970) for the cotton plant. If we are concerned only with population levels of one or more species which are confined to a particular part of the host then, obviously, our sample unit can be restricted to that plant part. For example, in assessing spider mite and/or predaceous mite populations on apple trees, a predetermined number of leaves is collected on each sample date and counts made to determine the average number present per leaf. Subsequent similar samples yield population trends.

Sample Interval

Determining proper sampling intervals (i.e., the frequency at which samples are to be taken) requires consideration of such factors as stage and growth rate of the crop, generation time and reproductive capacity of the target arthropods, climatic factors which may influence overwintering and diapause, the purpose of the study, etc. Obviously, for pests with short generation times and high reproductive capacities like aphids, a short sample interval is necessary during the growing season if populations are to be effectively monitored. For many crop pests, sampling at weekly intervals during the growing season is a routine practice, and samples often are taken even more frequently when pest populations begin to approach the economic threshold. Sampling at short intervals (every 3 or 4 days) may be necessary in certain crops such as vegetables where economic threshold levels are very low, or where economic infestations develop rapidly.

On the other hand, with some pests which complete only one or two generations per year, it is sometimes possible to obtain data needed for pest management decisions by sampling for a particular stage only once or a few times each year. Sampling to determine the density of overwintering pupae or eggs of certain forest pests, or for grasshopper egg pods per unit area in early spring are examples.

Sample Size

Determining optimum sample size (i.e., the number of sample units needed to provide data necessary for confident decision making) is particularly critical and sometimes difficult. If too few sample units are employed, statistical analyses are inconclusive and no confident decisions can be made. If more sample units are collected than are needed to reach a decision, time and money are wasted in collecting and processing excess data. Frequently, more or less standardized procedures have been developed for particular pests of a given crop. Watson *et al.* (1976) cite several such procedures. For example, in average-size fields of crops such as cotton, it is common practice to take samples in four areas, more or less equidistant from the center of the field and from each other, but well removed from the sides and ends of the field. Any unusual activity anywhere in the field is also noted. Some crops, such as alfalfa and certain vegetables, are sampled as the sampler moves diagonally across the field. Nielson (1957) found that the most practical and efficient method of sampling spotted alfalfa aphid populations was a leaf-count method in which the sampler counted the total number of aphids on three trifoliate leaves, each selected at random from the top, middle and bottom of the alfalfa plant. He recommended replicating each such sample unit 10 times at each random interval of 20 to 30 walking steps, while proceeding diagonally across the field.

Sequential sampling is an important sampling procedure which has been developed as a means of avoiding problems of insufficient and excessive sample sizes. Sequential sampling schemes have been developed for several important crop pests (e.g., Sylvester and Cox, 1961; Wolfenbarger and Darroch, 1965). Essentially, a sequential sampling scheme is one in which the sampler decides on the basis of the first samples taken whether or not further samples are required to reach a decision. If the number of pest individuals is very high in the first samples, the decision to treat is called for; if pest numbers are very low, a no-treatment decision is mandated. If the numbers fall within an intermediate range, further samples must be taken. After each sample the situation is re-appraised. Thus, sampling is continued until the preestablished upper or lower decision levels are attained, or until a predetermined number of samples has been taken, even if no decision is reached. In the latter case, the conservative decision (to treat) is often made. Waters (1955) reported that the reduction in sampling time obtained by the use of the sequential method often exceeds 50 percent. This procedure has been employed in alfalfa weevil pest management, worked out by IPM interstate research with contributions from W-84.

However, the establishment of a sequential sampling program requires substantial knowledge of the spatial distribution of the species to be sampled. It must be determined whether the distribution of the target species is most nearly random, clumped or regular, and whether the type of distribution varies with population density. These kinds of data, plus data on economic thresholds and the aid of a competent statistician, are prerequisites for the development of a satisfactory sequential sampling scheme. For a more detailed consideration of sequential sampling see Onsager (1976).

Sampling Methods and Devices

Careful comparison and selection of sampling methods and sampling units to be employed in a pest management project can result in substantial savings in both the time and the effort required to obtain reliable data. For example, Nielson (1957) made a statistical comparison of six methods used for sampling spotted alfalfa aphid populations: the sweep net; the Henderson fork method (in which a fork-like tool holding four glass slides covered with an adhesive material is held beneath the plant while the plant is shaken—the adhering aphids are then counted); and four different plant-inspection methods. He showed that a simple leaf-counting method was the most efficient and accurate of those tried.

Young and Tugwell (1975) compared different methods of sampling tarnished plant bug populations on cotton in Arkansas. The D-vac, sweep net, drop sheet and seven different plant inspection methods were compared in three different plot sizes. The D-vac, sweep net and terminal inspection methods sampled only the upper 10 to 12 inches of the plants, whereas the drop sheet and square inspection methods sampled entire plants. These authors concluded that the sweep-net method was the most rapid by 8 to 12 minutes per 25-acre field. The drop-sheet method produced the most bugs per minute of effort, although significantly fewer adults per minute were collected by that method than with the D-vac or sweeping net. The reliability of the mean population estimates, as indicated by coefficients of variation, was greatest with the drop-sheet method. When all tests were considered, the drop-sheet method appeared to be the most efficient, and the terminal-inspection method the least efficient, of the methods compared.

In sampling populations of active predators such as *Nabis, Geocoris,* and the like, several workers have found that whole plant samples give more reliable data than do the more usual types of stratified samples in which sampling is restricted to a portion of the plant or to particular plant parts. (Anonymous, 1955; Leigh *et al.,* 1970).

Inspection of plants or plant parts is a method widely used to sample insect populations or their damage. Generally a sample unit appropriate to the crop and insect involved is selected. Observational samples may prove adequate for certain comparative assessments. In taking such samples, the number of a particular insect observed over a period of time or within a unit area, or both, is recorded. This method, and the sweep-net method, have been used extensively in assessing certain predator populations.

The standard 15-inch-diameter sweep net has long been used in agricultural entomology to sample insect populations. There are drawbacks to the use of this method; realizing these drawbacks, however, it does provide a convenient, practical and generally rapid method of obtaining usable population census data. Rakickas and Watson (1974) utilized the method to determine population trends of *Lygus* spp. and seven important predator groups in strip-cut alfalfa. An important factor in the reliability of this method is the standardization of the sweep for different individuals involved in the sampling.

Michelbacher and Smith (1943) used the sweep net effectively in alfalfa as a method to determine if populations of the alfalfa caterpillar, *Colias eurytheme* (Boisduval), required control with insecticides. Their research determined that if no more than 10 *Colias* larvae per sweep could reach maturity (i.e., not parasitized) then insecticidal treatment was not warranted. This provided a basis for the development of supervised control of alfalfa pests.

Cothran and Summers (1972) found that the sweep-net method was unsatisfactory for sampling total alfalfa weevil populations. Young weevil larvae tended to burrow into the terminal buds, and therefore were seldom dislodged by the net. This feeding behavior lasts for several weeks during the spring. During this period larval populations may approach maximum levels without appearing in sweep-net counts. These authors found that the square-foot sample count, employing a randomly dropped 1-foot-square metal frame and then dissecting the tips of all stems within the frame (plus extracting the litter and top ½-inch of soil to recover older larvae), was a much more accurate method for monitoring alfalfa weevil populations. Their findings demonstrated advantages of using an absolute sampling unit as compared to a relative one. Ruesink and Kogan (1975) pointed out that some insects tend to drop to the ground when a shadow passes over them. Such species cannot be sampled adequately with a sweep net unless the sweeper walks toward the sun.

The development of satisfactory sampling methods and procedures usually require a thorough knowledge of the biology and ecology of the species involved. For example, Leeper (1974) found that *Lixophaga sphenophori* (Villeneuve), a parasite of the sugarcane weevil, *Rhabdoscelus obscurus* (Boisduval), could be sampled quite easily, with a sweep net or by using glass vials, after spraying the foliage of food-source plants with a honey-water mixture which served as an attractant. Observations made by Leeper on important population attributes such as sex ratio and density within sugarcane fields as compared to that along the field margins led to a more precise trap sampling method developed by Topham and Beardsley (1975).

The development of an adequate sampling method and determination of an economic threshold for the pink bollworm, *Pectinophora gossypiella* (Saunders), helped reduce insecticide use on cotton in Arizona. Watson and Fullerton (1969) determined that no yield losses occurred in central Arizona where no more than 20 percent of the vulnerable bolls were infested on any sample date. Slosser and Watson (1972*a*) showed that the initial population developed on early squares, with the next generation developing on the bolls on the lower half of the plants. Subsequent generations developed on bolls higher on the plant. The most susceptible bolls were 10 to 20 days old. Infestation levels are similar within the field and at the edge (Slosser and Watson, 1973) and there is a correlation between the numbers of pink bollworm larvae per boll and the percent boll infestation (Slosser and Watson, 1972*b*). This means that the intensity of the infestation can be predicted. Weekly samples of bolls 10 to 20 days of age, collected at random within a cotton field can be used to determine economic infestations of pink bollworms, and can serve as the basis for insecticide applications.

The suction collection machine (D-vac, etc.) is a mechanical device now widely used for sampling arthropod populations in many crop situations. Operating on the same basic principle as the household vacuum cleaner, suction machines will pick up the majority of arthropods which do not adhere too tightly to host surfaces (Dietrick et al., 1959). A variety of hand-carried, back-pack mounted, and tractor or jeep-drawn models has been developed. Collection by these machines is generally rapid and efficient. The disadvantages of this type of sample-collection equipment are the relatively high initial cost and upkeep, and the fact that the method is relatively nonselective, often yielding samples which are highly contaminated with plant debris and with non-target species such as spiders, honeybees, and the like. Dietrick *et al.* (1959) used a suction machine to collect alfalfa arthropod samples from 1 square-foot sample units. Each sample was immediately refrigerated to reduce arthropod activity. Later, arthropods were extracted into alcohol, using a battery of modified Berlese funnels. Leeper and Beardsley (1976) found the back-pack mounted D-vac machine to be an excellent tool for sampling adult populations of the Albizzia psyllid and psyllid predators on *Acacia* spp. in Hawaii. This method was not effective for sampling the sessile immature stages of the psyllid, however. The latter were sampled by randomly collecting ten individual twig terminals from each sampling site into plastic bags, for later counting of psyllid nymphs and eggs in the laboratory.

Special methods have been developed for sampling populations of phytophagous mites and associated predators. One such method widely used is the leaf disc sample. A special punch is used to remove a 1-inch-diameter circular section from each leaf sampled. These sample units are isolated for later counting in the laboratory. For certain kinds of leaves which are relatively smooth and not too fragile, leaf-brushing machines have been developed which remove mites and distribute them on an adhesive-covered revolving glass disc for counting (Henderson and McBurnie, 1943).

Other methods for removing mites, mite eggs, and other small arthropods from plant material by means of various solutions, or by centrifuging, have also been developed. Venables and Dennys (1941) developed a method for estimating mite population densities, particularly in heavy infestations, whereby sample leaves were placed between sheets of white mimeograph paper and passed through the rollers of an ordinary clothes wringer. The crushed mites from both sides of the sample leaves left characteristic stains on the paper which could then be counted.

Trapping insects with mechanical devices of various types is one of the principal types of sample collection methods employed in pest management sampling and monitoring programs. Trapping devices range from simple and inexpensive to relatively complex and expensive. The term "trapping" implies that the device is exposed over a period of time during which it is usually unattended. Arthropods are trapped in the course of their normal activity, often in response to some attractive principle incorporated into the trap. Therefore, the results obtained from virtually all insect-trapping devices are strongly influenced by the various ambient environmental factors which affect insect activity, such as temperature, humidity, precipitation, wind, light intensity and the like. It is important to bear this fact in mind when making comparisons based on data obtained by trapping methods.

Among the simpler types of traps, the pitfall trap has found widespread application in monitoring populations of largely ground-inhabiting species, such as ants, elaterid beetles, carabid beetles, Collembola, etc. A simple container, such as a 100 ml widemouth glass jar, containing a solution of 3 percent formalin plus a small amount of detergent, and buried to the level of the surface of the soil, serves as the satisfactory trap in most situations. Populations of ants and other ground-inhabiting arthropods in Hawaiian pineapple fields have been successfully monitored by means of such traps placed at 40-foot intervals between alternate plant beds. Traps were collected and emptied at weekly intervals, except during periods of heavy rain when more frequent collection was necessary (Su and Beardsley, unpublished). Pitfall traps baited with attractants such as stale beer have been used by some workers. Another simple trapping device developed for relatively rapid survey of ant populations in pineapple fields involved the use of a food attractant (1:1 honey and water solution) applied to one end of standard wooden tongue depressors. These were placed upright in the soil at 20-foot intervals between planting beds. The trap sticks were re-examined after 3 to 6 hours (preferably at dusk or in the early evening) and the number of ants present on each stick estimated. By this method it was possible to map the approximate distribution of ant colonies throughout a 50-acre experimental pineapple planting in a relatively short time (4 man-days; McEwen and Beardsley, unpublished).

Light traps are widely used as insect survey and detection devices and to some extent for direct population supression. Much detailed information on the construction and use of light traps has been published (Frost, 1952; Nelson, 1967; Hienton, 1974). All light traps involve a light source for attraction, and a catching device coupled with a killing agent to effect rapid mortality. A variety of light sources has been utilized, but most traps now employ light in the near ultra-violet wavelengths to which most night-flying insects respond well. The type of collection device and killing agent used depend upon the purposes for which the trap is employed, and on the target species. Some types are equipped with electrocuting devices placed over collecting bags or troughs. Where specimens are to be preserved for identification and counting, killing chambers employing fast-acting poisons such as cyanide or ethyl acetate are frequently used. Isopropyl alcohol has been used as a collection medium for light traps by some workers, with good success (AliNaizee and Stafford, 1972*a*).

Light traps attract a wide range of insect species. They have been used most successfully in connection with survey and population sampling studies of lepidopterous pests such as the codling moth, the pink bollworm, the corn earworm and various kinds of hornworms, armyworms, and cutworms. They have also been useful in studies on certain other pests, such as scarab beetles and wireworms, the adults of which are nocturnal fliers.

Where direct comparisons have been possible, light traps have been found to be less sensitive as population-monitoring devices than are traps employing chemical sex attractants or virgin females. In a study on the omnivorous leafroller, *Platynota stultana,* Barnes and Busk, AliNiazee and Stafford (1972*b*), found virgin female traps to be six to seven times as effective as light traps in attracting adult moths. Light traps usually catch both sexes of the target species and this may provide an advantage over pheromone traps in some situations.

A major drawback in the use of light traps in programs involving one or a few target species is their general lack of specificity—light trap catches often contain large numbers of non-target species. Mechanical-sorting devices which separate trapped insects into different size classes can sometimes be used to reduce the amount of labor required for sorting out the target species.

Some insects are attracted to particular colors or shapes. Aphids and leafhoppers, for example, frequently are attracted to yellow or greenish-yellow surfaces, and *Rhagoletis* flies were found to respond positively to fluorescent Saturn-yellow rectangles and spheres (Boller and Prokopy, 1976; AliNaizee, unpublished). Sticky traps incorporating such attractant colors and shapes are commonly used for monitoring populations of aphids, leafhoppers and fruit flies.

Another type of trap sometimes used in sampling and monitoring insect populations is the malaise trap. This is essentially a tent made of fine netting, open on either two or four sides, into which insects fly or crawl. As most insects tend to crawl upward once they are on the net, they can be trapped in collecting devices positioned at elevated corners, ends, or at the apex of the tent (depending upon the trap design). In most models, flaps attached at the sides and top partially close the openings, and reduce the chance of escape. Malaise traps are most effective in trapping actively flying insects. Various kinds of suction traps, rotary nets, and windsock nets also have been designed for sampling populations of flying insects.

Various types of ground-emergence cages have been used to trap newly emerging ground-inhabiting insects such as beetles, Lepidoptera and Diptera which pupate in the soil. Workers in different areas have constructed traps to suit their own needs. Basically, these involve enclosing a fixed area with a cloth or screen cage, and counting the number of insects emerging from the enclosed piece of ground. For overwintering forms, such cages provide a definite indication of date of emergence of first insects in a particular area, as well as a relative indication of population size. For example, the ground emergence cages used for the western cherry fruit fly, *Rhagoletis indifferens* Curran, are 3 inches x 3 inches pyramid-shaped cages with plastic vials mounted on the top (AliNiazee, 1974). The flies, being positively phototrophic, move to the top part of the cage and into a vial soon after emergence. The vials are checked daily to determine the peak of emergence. A count of total number of flies per cage gives a relative indication of population size at a particular location. However, since the population size is dependent on so many variables, no accurate estimates of total populations can be obtained from a few randomly-distributed cages. To produce valid population estimates, large numbers of cages would be required.

Many types of insect traps employ attractants which depend upon olfactory responses of the target species. Some attractants are food substances such as honey or other sugar sources for ants, or protein hydrolysate which is attractive to many insects including adults of tephritid fruit flies. Many other insect food substances, or volatile compounds normally associated with food substances, have been used to trap insects, both for survey and population reduction purposes. For example, containers of terpinyl acetate have been widely used to trap fruit crop insects, such as the Oriental fruit moth, *Grapholitha molesta* (Busk), and the codling moth, *Laspeyresia pomonella* (L.). Ammonium carbonate is a general attractant for *Rhagoletis* species and has been used extensively in traps for these fruit flies.

Other chemicals which are sometimes used in insect traps are species and sex specific lures; for example, methyl eugenol, which is highly attractive to males of the Oriental fruit fly, *Dacus dorsalis* Hendel, does not attract such species as the melon fly or the Mediterranean fruit fly.

Various kinds of traps baited with food or hosts have been devised to sample populations of beneficial parasites and predators. For example, Topham and Beardsley (1975) used perforated plastic vials each containing a host larva (the New Guinea sugar cane weevil, *Rhabdoscelus obscurus* [Boisduval]) in a macerated coconut husk medium to obtain data on the abundance and distribution of the parasitic tachinid, *Lixophaga sphenophori* (Villeneuve), in Hawaiian sugar cane fields. Host vials were distributed on cane plants at 50-foot intervals along transects through the fields, and after 7 days were retrieved and held for parasite emergence. To obtain data on field distribution of the two sexes of this parasite, these authors used cylindrical traps fabricated from aluminum window screen and closed at bottom and top. A sheet of thin plastic was taped to the outside of the screen cylinder and coated with an adhesive mixture to serve as a trapping surface. Traps were baited by placing 2-week-old *R. obscurus* grubs in sections of sugar cane and pieces of cane leaves sprayed with a honey-water solution inside the trap, thus utilizing both food and hosts as attractants.

The pheromone trap is a relatively new development in insect-trapping techniques. Pheromones are specific chemicals produced by insects which attract mates, or, in social species, are used to mark trails, elicit alarm responses, etc. Sex pheromones are generally quite species specific. For certain pest species, where it has been possible to isolate and synthesize these chemicals, highly efficient detection and survey traps have been developed (Jacobson, 1972; Shorey, 1974).

Some sex pheromones are available commercially for trapping or for direct suppression of insect populations. Sex pheromones of some important insect pests including the European corn borer, codling moth, cabbage looper, pink bollworm, gypsy moth, bark beetles, cotton boll weevil, Indian meal moth, Oriental fruit moth and various leafroller species have been synthesized. Several companies produce sex pheromone traps for several pest species, and interested growers can obtain complete trapping kits.

Various types of sex pheromone traps have been constructed by different workers to suit their needs (Jacobson, 1972). Standard sticky traps made out of ice cream cartons, with a rubber septum impregnated with a sex pheromone, have been widely used in this country. Pheromone traps usually employ a pheromone carrier which releases the pheromone at a constant rate, and a sticky surface to trap attracted insects.

In spite of recent advances in the pheromone research work, there are still many insects for which either the existence of a pheromone is unknown, or if known, the structure of the pheromone has not been identified. In some instances, virgin-female traps employing unmated females as pheromone sources have been used. Sometimes such traps are more effective than synthetic pheromone traps, but in other instances the synthesized pheromones were more effective (AliNiazee, 1976). Virgin-female traps are more difficult to keep in operation because they require a sequential supply of virgin females.

A number of factors affect the performance of virgin female or synthetic pheromone traps; for example, prevailing weather conditions, trap location, host plants, physiological conditions of males and females, the potency and rate of emittance of the pheromone from the source, and the type of trap used. Therefore, extensive biological work is needed to determine the optimum utilization of these traps in insect population sampling and monitoring programs, and a proper interpretation of the field data is required. For a detailed discussion of pheromone traps see Shorey *et al.* (1968), Jacobson (1972), and Shorey (1974, 1976).

Damage Indices

Crop pests are generally classified as either direct pests (those which directly damage the marketable product) or indirect pests (those which damage parts of crop plants which are not marketed). With direct pests, a relatively small amount of damage may result in great economic loss. With indirect pests, greater numbers generally can be tolerated before yields are affected to the point where economic losses result.

Damage indices for direct pests are based directly on examination and ranking of the damage to the marketable product, such as "infested" vs "non-infested" in cases where any observable damage results in the product being discarded (e.g. codling moth damage to apples), or into various classes of damage in crops where damage causes devaluation or down-grading of the product but not complete rejection. In general, damage indices for direct pests are much more easily correlated with economic threshold levels than is damage by indirect pests (such as defoliators) where the economic effect of damage may vary widely with season, age of the crop, crop variety, etc. For these reasons, correlations between indirect damage and economic loss may only be valid under the specific conditions under which they were developed.

Damage indices for indirect pests frequently involve defoliators or root feeders, and sampling is used to develop estimates of yield reduction based upon the degree of defoliation or root loss. For defoliation, visual estimation is a common method. Obviously, individual estimators can vary widely in their accuracy. When greater accuracy is required than is normally attainable by visual estimation, more precise methods must be employed. Planimeters and grids have been used by some workers to measure area of leaf loss due to defoliators. Various kinds of photometric devices (photoplanimeters) which measure total intact leaf area also have been developed. By means of such a device the average area of a sample of damaged leaves can be compared with that of undamaged leaves to estimate total leaf loss due to defoliators.

Measuring damage to root systems is generally more tedious, often requiring detailed examination of sample roots. An indirect approach to this problem is the "pull technique" by which the force necessary to uproot sample plants is measured with a special device (recording dynamometer). These values can then be compared with a set of known values which correlate visually determined root damage with force necessary to uproot. However, differences in soil type and soil moisture can introduce a large element of error into such correlations (Ruesink and Kogan, 1975).

Literature Cited

AliNiazee, M. T.
1974 The western cherry fruit fly, *Rhagoletis indifferens*. 1. Distribution of the overwintering pupae in the soil. Can. Entomol. 106: 909-12.
1976 Field studies on sex pheromone trapping of the filbert leafroller, *Archips rosanus* in Oregon. Ann. Entomol. Soc. Amer. 69: 821-24.
AliNiazee, M. T. and E. M. Stafford.
1972*a* Seasonal flight patterns of the omnivorous leafroller and grape leaffolder in central California vineyards as determined by black light traps. Environ. Entomol. 1: 65-68.
1972*b* Virgin female traps aid control survey for omnivorous leafroller in the San Joaquin Valley vineyards. Calif. Agric. 26: 5-6.
Anonymous.
1955 Survey methods for predators on cotton. USDA Coop. Econ. Insect Report. 48 : 5.
Boller, E. F. and R. T. Prokopy.
1976 Bionomics and management of *Rhagoletis*. Ann. Rev. Entomol. 21: 223-45.
Cochran, W. G.
1953 Sampling Techniques. John Wiley & Sons, Inc. New York, N.Y., 330 p.
Cothran, W. R. and C. G. Summers.
1972 Sampling for Egyptian alfalfa weevil: a comment on the sweep net method. J. Econ. Entomol. 65: 689-91.
Dietrick, E. J., E. I. Schlinger and R. van den Bosch.
1959 A new method for sampling arthropods using a suction collection machine and modified Berlese funnel separator. J. Econ. Entomol. 52: 1085-91.
Dumas, R. A., W. P. Boyer and W. H. Whitcomb.
1962 Effect of time of day on surveys of predacious insects in field crops. Florida Entomol. 45: 121-28.
Finney, D. J.
1953 An Introduction to Statistical Science in Agriculture, Oliver and Boyd, London, England. 179 pp.
Frost, S. W.
1952 Light traps for insect collection, survey and control. Penn. Agric. Exp. Sta. Bul. 550.

GONZALEZ, D.
1971 Sampling as a basis for pest management strategies. Proc. Tall Timbers Conference on Ecological Animal Control by Habitat Management (Tall Timbers Research Station, Tallahassee, Florida). No. 2 : 83-101.
HANSEN, M. H., W. N. HURWITZ, and W. G. MADOW.
1953 Sample Survey Methods and Theory, a. John Wiley & Sons, Inc., New York, NY., 638 pp.
HENDERSON, C. F. and H. V. McBURNIE.
1943 Sampling technique for determining populations of the citrus red mite and its predators. USDA Circular 671, 11 pp.
HIENTON, T. E.
1974 Summary of investigations of electric insect traps. USDA, ARS Tech. Bul. 1498, 136 pp.
JACOBSON, M.
1972 Insect sex pheromones. Academic Press, N.Y. 382 pp.
KUEHL, R. O. and R. E. FYE.
1972 An analysis of the sampling distribution of cotton insects in Arizona. J. Econ. Entomol. 65: 855-60.
LEEPER, J. R.
1974 Adult feeding behavior of *Lixophaga sphenophori*, a tachinid parasite of the New Guinea sugarcane weevil. Proc. Hawaiian Entomol. Soc. 21: 403-12.
LEEPER, J. R. and J. W. BEARDSLEY.
1976 The biological control of *Psylla uncatoides* (Ferris and Klyver) (Homoptera:Psyllidae) on Hawaii. Proc. Hawaii. Entomol. Soc. 22: 307-21.
LEIGH, T. F., D. GONZALES and R. van den BOSCH.
1970 A sampling device for estimating absolute insect populations on cotton. J. Econ. Entomol. 63: 1704-06.
LINCOLN, C., G. C. DOWELL, W. P. BOYER, and R. C. HUNTER.
1963 The point sample method of scouting for boll weevil. Arkansas Agric. Expt. Sta. Bul. 666. 31 pp.
MICHELBACHER, A. E., and R. F. SMITH.
1943 Some natural factors limiting the abundance of the alfalfa butterfly. Hilgardia 15: 369-97.
MORRIS, R. F.
1960 Sampling insect populations. Ann. Rev. Entomol. 5: 243-64.
NELSON, S. O.
1967 Electromagnetic energy. Pages 89-145. *In* Kilgore and Doutt (Editors). Pest Control. Academic Press. New York.
NIELSON, M. W.
1957 Sampling technique studies on the spotted alfalfa aphid. J. Econ. Entomol. 50: 385-87.
ONSAGER, J. A.
1976 The rationale of sequential sampling, with emphasis on its use in pest management. USDA Tech. Bul. 1562, 19 pp.
RAKICKAS, R. J. and T. F. WATSON.
1974 Population trends of *Lygus* spp. and selected predators in strip-cut alfalfa. Environ. Entomol. 3: 781-84.
RUESINK, W. G. and M. KOGAN.
1975 The quantitative basis of pest management: sampling and measuring. pp. 309-51. *In* R. L. Metcalf and W. Luckman (Editors) Introduction to Pest Management. John Wiley and Sons, New York.
SHOREY, H. H.
1974 Environmental and physiological control of insect sex pheromone behavior. Pages 62-80 *In* M. C. Birch (Editor). Pheromones. North Holland Publication Co., Amsterdam.
1976 Animal Communication by Pheromones. Academic Press. New York, N.Y. 167 pp.
SHOREY, H. H., L. K. GASTON, and R. N. JEFFERSON.
1968 Insect sex pheromones. Adv. Pest Control Res. 8: 57-126.
SLOSSER, J. E. and T. F. WATSON.
1972*a*. Population growth of the pink bollworm. Ariz. Agric. Expt. Sta. Tech. Bul. 195. pp. 32.
1972*b* Relationship between percent boll infestation and mean number of pink bollworm larvae per boll. J. Econ. Entomol. 65: 1493-95.
1973 Comparison of pink bollworm infestation levels between various portions of a cotton field. J. Econ. Entomol. 66: 278-80.
SOUTHWOOD, T. R. F.
1966 Ecological Methods with Particular References to the Study of Insect Populations. Methuen, London. 391 pp.
SUMMERS, F. M.
1966 The oriental fruit moth in California. Calif. Agric. Expt. Sta. Circ. 539. 18 pp.
SYLVESTER, E. S. and E. L. COX.
1961 Sequential plans for sampling aphids on sugar beets in Kern County, California. J. Econ. Entomol. 54: 1080-84.
TOPHAM, Marshall and J. W. BEARDSLEY.
1975 Influence of nectar source plants on the New Guinea sugarcane weevil parasite, *Lixophaga sphenophori* (Villeneuve). Proc. Hawaiian Entomol. Soc. 22: 6 pp.
VENABLES, E. P. and A. A. DENNYS.
1941 A new method for counting orchard mites. J. Econ. Entomol. 34: 324.
WATERS, W. E.
1955 Sequential sampling in forest insect surveys. Forest Sci. 1: 68-79.
WATSON, T. F. and D. G. FULLERTON.
1969 Timing of insecticidal applications for control of the pink bollworm. J. Econ. Entomol. 62: 682-85.
WATSON, T. F., L. MOORE and G. W. WARE.
1976. Practical Pest Management. W. H. Freeman, San Francisco. 196 pp.
WOLFENBARGER, D. A. and J. G. DARROCH.
1965 A sequential sampling plan for determining the status of corn earworm control in sweet corn. J. Econ. Entomol. 58: 651-54.
YOUNG, S. C. and P. TUGWELL.
1975 Different methods of sampling for clouded and tarnished plant bugs in Arkansas cotton fields. Univ. of Arkansas Agric. Expt. Sta. Report Series 219. 12 pp.

Chapter 3.

ECONOMIC THRESHOLDS AND ECONOMIC INJURY LEVELS

S.C. Hoyt and R.G. Simpson

Introduction and Importance

The use of modern insecticides to control insect pests has been a necessary part of crop production and the prevention of insect-related health problems. In all probability, insecticides will continue to be used and needed for some time to come. However, as Geier (1966) points out, pest control measures amount to hardly more than bulldozing nature without regard to the consequences. For too long a time many growers, extension workers, chemical representatives, and entomologists have used pesticides either in high dosages, as extra insurance, or on a calendar basis. These practices have brought about adverse biological and ecological impacts and often compounded the original problem.

Control measures should not be carried out unless it is known that the pest is present in sufficient numbers to cause an economic loss. This statement is made with the assumption that adequate research has been done with the particular pest and its host to establish the economic threshold. Treatments applied below this threshold, where the cost of control exceeds the value of the crop being protected, result in financial loss. All too often pesticide treatments have been made on a "visibility threshold," which implies that if insects are observed even in trace or low numbers chemical controls are necessary. Recommendations frequently call for treatments when buds are open, or when heads begin to form, leaving nothing to the discretion of the grower or consultant. It is evident that economic injury levels and economic thresholds for most pests on a majority of crops are not available to guide those individuals making recommendations. The reader at this point will wonder why they have not been established. The answer is that it is a complicated process to establish either level. Only about 12 economic thresholds have been established on ten important crops for specific insects in California (Stern, 1973) but in this regard California is far ahead of most other states. Many of the biological and ecological complexities in the determination of economic thresholds and economic injury levels will be discussed in the following sections.

Definitions

Although there are different terms expressing essentially the same meanings, those defined by Stern (1973) seem to be the most widely accepted and are presented here. *Economic threshold* means the density at which control measures should be determined to prevent an increasing pest population from reaching the economic injury level. *Economic injury level* is defined as the lowest population density that will cause economic damage. Judging by fragmentary evidence these levels are almost invariably higher than expected (Chant, 1966). In a pest management program, the economic injury level must first be established in order to employ the concept of economic threshold. The threshold is the pest-density level at which control applications should be made. Although considerable disagreement exists as to whether the terms "integrated control" and "pest management systems" are synonymous, the present emphasis and popularity of these programs reflect a direct dependency upon continued research to determine the economic thresholds and injury levels for the crops and insects on which either system is acting. The lack of soundly based economic thresholds and injury levels is well verified by a survey conducted by Stern (1973) and further stated by DeBach (1974) and Hoyt and Burts (1974).

Factors Influencing Economic Injury Levels

Crop condition

The fact that economic injury levels vary for a given species of insect on a particular crop is often overlooked. The presence of insects on a crop is not necessarily a threat of economic injury—pest populations can be tolerated at levels that have little or no effect on crop yields and quality. Fluctuating injury levels need to be quantified for the state of plant development, time of year, plant variety, cropping practices and the purpose for which the crop is intended. A crop that has been sprayed frequently will not have the same economic threshold as the identical crop that has been sprayed only occasionally or left unsprayed. This has been attributed to the elimination of natural enemies (DeBach, 1974).

It is obvious that a plant or tree that is maintained in a vigorous, healthy condition by proper irrigation, a well-balanced fertilizer schedule, and other good cultural practices (see "Cultural Practices in Pest Management") will be able to tolerate a greater number of pests than one in a less vigorous condition.

Plant varieties resistant to insect attacks have much to offer as a method of limiting insect damage and the level of injury. When available, resistant varieties of plants cost little or no more than susceptible varieties. Plant resistance is not necessarily a permanent solution, however, because insects may respond to selection pressures by developing a biotype that successfully attacks the resistant varieties. Most resistant plants are those of field crops such as wheat, corn, and alfalfa varieties.

These conditions and others modify the economic threshold and injury levels of a given pest on a particular plant.

Economics

The economic variables that enter the picture of pest control and the concept of economic injury levels are complex. Due to the large number of factors, the threshold and injury levels will change continually with fluctuating economic and environmental conditions. A decision to treat with an insecticide must be determined by the cost of the application balanced against the increased value of the crop that can be protected. Crop values can be affected by market conditions on a regional, national or international basis.

There are cases where overproduction by increased yields or plantings create low market prices and compel growers to dump crops (peaches in Colorado) or to plow them under (wheat in Kansas and carrots in Colorado). In these instances, the potential insect injury is not an economic consideration.

The non-edible portions of many plants can tolerate moderate losses before the crop yield is affected. Alfalfa growers frequently lose slight tonnages to leaf-eating alfalfa weevils without appreciable dollar losses because the bulk of the crop weight is in the stems and thus unaffected by weevils. It has been proposed that in the future, alfalfa may be purchased from growers on the basis of the protein content which essentially is in the leaves—if and when that occurs, the grower will be concerned with economic injury by alfalfa weevils at a much lower level than at the present time.

It is widely known that some insecticides may bring about conditions favorable for rapid buildup of pest insects which creates a need for more treatments. Frequently this buildup can be attributed to resistance by the pest resulting from continual use of one insecticide or a particular group of insecticides. Also, there may be a buildup of secondary pests following treatments which may require additional sprays (Stern, 1973). These situations can add greatly to the total cost of production. An interesting concept was presented by Croft and Brown (1975) who feel that a more relaxed (higher) economic threshold would permit additional hosts (pests) to survive, and that this would allow natural enemies to multiply and respond by becoming resistant to a wider spectrum of insecticides. Other workers have expressed the reverse idea, that reduced selection pressure delays resistance development in pests. These examples are but a few of the complexities that need to be considered in the economics of determining fluctuating economic thresholds.

Climate

A physical factor such as climate can alter economic thresholds by favoring or by not favoring an insect species, or can alter economic injury levels by modifying the susceptibility of plants to attack by insects. In natural control, climatic factors are perhaps the most important. It has long been known that certain pests (such as the brown wheat mite on wheat, twospotted mites on corn, grasshoppers on rangeland and cultivated crops, spotted alfalfa aphids on alfalfa) increase their numbers dramatically in periods of drought. Plants, if dependent solely upon rainfall, are often stunted and suffer not only from the lack of moisture but also from the increased numbers of insect attacks. Drought conditions on non-irrigated tree fruits may lead to explosive outbreaks of mite populations and to a greatly reduced tolerance for mite damage. Conversely, pea aphids on alfalfa increase many-fold in cool, wet weather, although there is little apparent injury to the plants. Populations of European corn borers decline during succeeding cool wet springs and evidently are more subject to attack by parasites.

Legal factors and public attitudes

Quarantines against the shipment of commodities containing even a single specimen of certain pest species reduce the tolerance for that species to an extremely low level. Similarly, regulations governing the presence of insect parts in processed foods may have a substantial impact on economic injury levels and treatment thresholds. The buyers or consumers of agricultural produce may also play a part in establishing economic injury levels by refusing to purchase items with minor blemishes or other evidence of the presence of insects. These blemishes can be equated to serious damage for fresh market consideration if consumers refuse to accept them.

All of the above tend to reduce the tolerance for the presence of the pest, regardless of whether it has caused "damage" to the consumable product in many cases. In complying with regulations and consumer demands, extensive pesticide programs and/or hand sorting may be required. This may, in effect, reduce yield, increase production costs and decrease the amount of natural enemies present on the crop.

If environmental improvement and long-term stable pest damage control through pest management are valid goals, then it seems important to question measures which run counter to these goals. Quarantines which are unnecessary or ineffective, prohibition of insect parts which do not constitute a health hazard or reduce the palatability of processed foods, and demand for nearly-perfect produce in markets place extreme stresses on agroecosystems. The solutions to these problems are political or educational rather than biological.

Pest-related factors

One of the main pest-related factors which influences economic injury levels is the feeding habits of the pest species. The tolerance for numbers of a species which attacks the consummable part of the crop (thus directly affecting yield or quality) will generally be low. This is particularly true for pests of fruits and vegetables, but less so for grains or forage crops. With species which cause indirect damage by feeding on nonconsumable plant parts, the plant can often tolerate moderate to heavy populations. A small number of some species, however, may result in death of plants or major loss through girdling of stems or tree trunks, or through the transmission of pathogens.

The time of year that a pest is prevalent in relation to crop development can also be an important factor in the economic injury level and the economic threshold. Stern (1973) discussed economic thresholds for lygus bugs on cotton, indicating the need for treatment when 10 or more bugs per 50 sweeps occurred on 2 consecutive sampling dates during prebloom to 2 weeks after bloom initiation, but 20 to 30 bugs per 50 sweeps could be tolerated after this early fruiting period. On apples, conservative thresholds for treatment of European red mite of 5 per leaf in May and early June, or 20 per leaf in late June or later are suggested.

The potential for increase of a pest population does not affect the economic injury level but may greatly influence the economic threshold. Factors which tend to dampen population increase and which are predictable, such as effective natural enemies, may allow the economic threshold to be increased. Anticipated periods of unusually favorable weather for increase in pest numbers may require a decrease in the numbers allowed before treatment is suggested.

Determining Economic Injury Levels

Even with direct pests, determining economic injury levels rarely is as simple as measuring the loss in yield, determining the value of the loss, and subtracting the costs of material and application of controls. There may be added costs of sorting damaged produce, and the control measures may have additional benefits through control of non-target pests or liabilities through destruction of natural enemies and other environmental impacts.

With indirect pests, it is important to compare different levels of infestation where cropping conditions are as uniform as possible. If plant ages and sizes, varieties, soil, slope, irrigation or other physical factors or cultural practices vary it may be difficult or impossible to determine the effect of the pests on yield or quality. Even under the best of conditions it may be difficult to obtain quantitative differences related to pest density. This does not mean that these variations should not be measured, as other stresses placed on plants may greatly influence their tolerance for a pest. The effects of these variables should, however, be measured between tests, not within them.

Measuring Pest Damage

In the most straightforward situations, it may only be necessary to establish different levels of the pest population and determine yield per unit area or per plant related to these levels. In some pest:crop interactions, it is much more time-consuming and difficult. With mites on apples, attempts have been made to relate density to total yield, fruit growth, color, internal chlorophyll, soluble solids, acids, firmness, sunburn and taste at harvest and after 3 and 6 months of storage. In the storage studies, internal breakdown, scald and bitter pit were also measured. Because a mite population can affect bloom and fruit-set the year following the outbreak, effect on yield in the second year was also measured. The most important of these measurements were fruit size and color, but subtle changes in the other factors can influence storage life or quality.

It may be necessary to measure effects of different densities at different times of the year or at different growth stages of the crop. Where the crop load may vary substantially from year to year it may be desirable to measure effects on plants with different crop loads.

Measuring Economic Loss Due to Pest Damage

Since complex marketing factors may result in wide fluctuations in the price of agricultural products each year, or even during the course of a marketing season, loss due to pest damage may also vary. It is even possible that moderate amounts of damage may result in no loss or in a gain in crop value on an area-wide basis. It is frequently necessary to make pest control decisions well in advance of the establishment of the value of the crop. Under these conditions accurate predications of crop size and price are useful in evaluating potential dollar loss.

Once a relationship between pest density and dollar loss is established it is necessary to add to this figure any added benefits from control measures, or subtract from it the cost of the control practice and any liabilities associated with it to determine the economic injury level of the pest. The economic threshold is then established by determining at what level control measures must be undertaken to prevent pest density from exceeding economic injury level.

In some cases it would take years of research to establish valid levels which take into account the many variables listed above. This is one reason why so few economic thresholds based on sound, quantitative data have been developed. In pest management programs it is important to establish thresholds, even if they are based on limited previous experience with the pest. These can be adjusted as additional experience or quantitative data are gained. The initial threshold is nearly always established at a conservative level, and as more information becomes available, the density at which control measures are recommended is increased.

Monitoring, Prediction and Economic Thresholds

An economic threshold may by itself have little meaning for practical purposes unless adequate methods of monitoring or sampling the pest population are used by the growers or pest management consultants. Additionally, the ability to predict pest density increases or decreases, peak populations, and the period of time a population will be present from a sample or series of samples is important to establishing a realistic relationship between the economic injury level and the economic threshold. The use of computer simulation models is providing a more sophisticated method of predicting the outcome of these complex interactions.

Literature Cited

CHANT, D. A.
1966. Integrated control systems. pp. 193-218. *In:* Scientific aspects of pest control. National Academy of Science. Washington, D.C. 470 p.

CROFT, B. A., and A. W. A. BROWN.
1975. Responses of arthropod natural enemies to insecticides. Ann. Rev. Entomol. 20: 285-335.

DEBACH, P.
1974. Biological control by natural enemies. Cambridge University Press, London. 323 p.

GEIER, P. W.
1966. Management of insect pests. Ann. Rev. Entomol. 11: 471-490.

HOYT, S. C., and E. C. BURTS.
1974. Integrated control of fruit pests. Ann. Rev. Entomol. 19: 231-52.

STERN, V. M.
1973. Economic thresholds. Ann. Rev. Entomol. 18: 259-80.

Chapter 4.

NATURAL ENEMIES AND PREDATOR-PREY RATIOS

K. S. Hagen and J. A. McMurtry

Introduction

Natural enemies of pests are usually present in most crops during the growing season, and they often suppress insect and spider mite populations. Whether or not the natural enemies suppress the pest populations before economic injury to the crop occurs depends, however, upon the abundance and activity of natural enemies in relation to the abundance of the pest and the economic injury level. Therefore, in order to establish effective predator-prey ratios, it is necessary to know at what population level a pest causes economic damage and how many natural enemies need be present to prevent the pest from reaching economic damaging numbers. The factors involved in determining economic injury levels of insects and spider mites are discussed in Chapter 3.

By taking several successive samples over time it can be determined if the pest population is increasing, remaining level or decreasing. This is the least amount of information a pest manager should determine before making any decision as to what to do about a potential pest. Ideally, the pest manager should also be monitoring (sampling) population levels of the pest's natural enemies. Not only should the pest manager be able to recognize the natural enemies, but he or she also must have some knowledge of their biologies and what biotic and physical factors may limit their activity. Only a few predator or parasite prey ratios have been determined.

Insect Predator-Prey Ratios

A few species of aphids, weevils and lepidopterous larvae have been evaluated as to the number of natural enemies necessary at different pest levels to bring about their control. Most examples discussed here deal with alfalfa pests as the economic injury level is known for most of these, and the biology and ecology of their important natural enemies are fairly well known.

Aphids

Natural enemies are important controlling agents of aphids (Hagen and van den Bosch, 1968; van Emden, 1972). Important aphid natural enemies include predators, parasites and pathogens.

Important common predators of aphids belong to the families Coccinellidae (lady beetles), Syrphidae (hover flies), and Chrysopidae (green lacewings).

Coccinellidae. In spring all common aphid-feeding lady beetles are active (Hagen, 1962, 1974). *Hippodamia, Coccinella, Olla* and *Adalia* are the common genera. However, during summer in the western U.S., the second generation of *Coccinella* species enters an obligatory dormancy (aestival diapause) in the adult stage; therefore, these *Coccinella* adults should not be relied upon to lay eggs at this time. Reproduction is usually necessary to obtain control of aphids. Most *Coccinella* species become active again in the fall but others do not. The common *Hippodamia (H. convergens* Guerin [fig. 1], *H. quinquesignata* Kirby, *H. sinuata* Mulsant and *H. parenthesis* [Say], *Olla abdominalis* [Say] and *Adalia bipunctata* [L.] can be reproductively active from spring to late fall providing there are ample aphids available and temperatures are high enough.

Hippodamia species can be distinguished from *Coccinella* species (fig. 2) by the tips of the femora projecting beyond the wing covers when viewed dorsally. Figure 3 shows a typical coccinellid larva.

Fig. 1. *Hippodamia convergens* Guerin.

Fig. 2. *Coccinella transversoguttata richardsoni* Brown.

Fig. 3. *Hippodamia quinquesignata punctulata* Leconte larva.

In alfalfa, the *Hippodamia* spp. which have arrived in the fields from overwintering sites in early spring (March in central California) require a minimum of either 100 large pea aphids (PA), 300 large spotted alfalfa aphid (SAA) or over 100 milligrams of mixed aphids and instars to lay eggs (Hagen and Sluss, 1966; see fig. 4). If egg laying occurs in early spring usually the resulting larvae will prevent economic damage to alfalfa by aphids during the spring. An average of one *Hippodamia* or *Coccinella* adult per one or two sweeps (5 square feet of alfalfa is sampled by one sweep using a 15-inch diameter net) requires a minimum average of about two large or ten small aphids per alfalfa stem in early spring with mean temperatures above 60°F (16°C) to produce and deposit eggs. In California, most overwintering adults die by the end of April. If egg laying has occurred by this time, there should be ample larvae present to stop aphid populations from causing damage, and the aphid populations (SAA, PA) will rarely exceed 30 per stem (Neuenschwander *et al.*, 1975; Smith and Hagen, 1966). An average of one *Hippodamia* or *Coccinella* larvae per sweep will control up to 50 aphids per stem in alfalfa stubble.

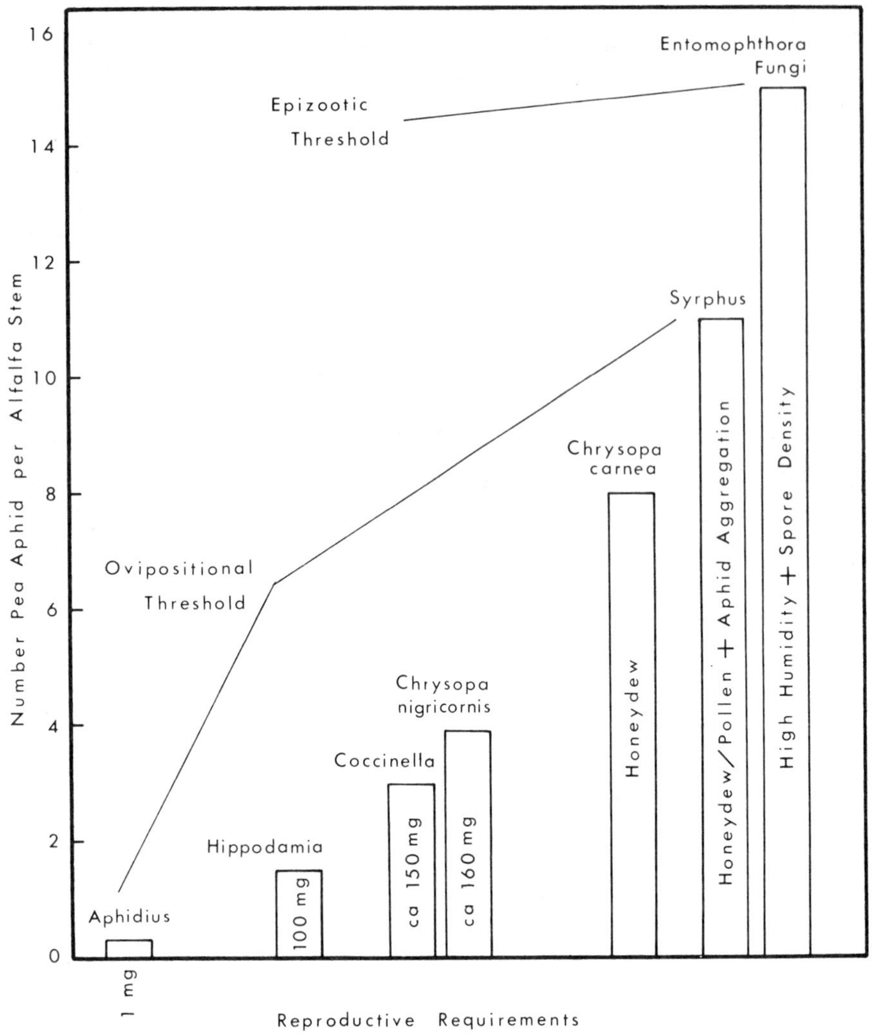

Fig. 4. Bars indicate the relative mean number of pea aphids on alfalfa stems, including leaves, necessary to induce oviposition in various aphid natural enemies. The information within the bars are minimum total biomasses of pea aphids or other requisites for the natural enemies to reproduce and for the fungus to cause an epizootic. One *Hippodamia* female, for example, must search about 50 stems to find enough aphids at 2 per stem to satisfy the 100 mg biomass of aphids needed for egg production and oviposition. (From Hagen, 1976.)

In late spring when the first generation coccinellid adults emerge, the beetle-aphid ratio changes. These second generation *Hippodamia* must consume over 130 large aphids (130 milligrams of aphids) in order to lay their first batch of eggs (Hagen and Sluss, 1966). Thus, one *Hippodamia* adult per sweep requires five large or fifteen small aphids per stem to induce oviposition. However, newly-emerging adults are so numerous that during May and June in California they usually destroy most of the few aphids that remain and leave the fields without ovipositing. Thus, the ratio of beetles to aphids is usually too high to permit egg production, but the aphid populations are generally reduced to such low levels by late May and June by adults that the aphids do not increase to high numbers again until August or September. In autumn when many of the coccinellids re-invade alfalfa fields, the same spring ratios may be used. However, if the temperatures become too cool in the fall, the aphids may continue to reproduce whereas the coccinellids will not because of differential low thresholds for reproduction between aphids and beetles, and diapause is induced in beetles under short days and low temperatures (18°C, 65°F).

Aphicide treatment levels for SAA are 40 per stem in the spring, 20 per stem in the summer, and for PA 40 to 50 per stem when alfalfa is less than 10 inches high, 70 to 80 when 15 inches high and 100 per stem when the alfalfa is more than 20 inches high (Bowen *et al.,* 1976).

Perhaps the above coccinellid adult-aphid ratios can be extrapolated to other low crops. One female beetle per sweep (5 square feet) in spring and fall should lay eggs if at least 500 mg of aphids are present in 5 square feet of crop.

Olla abdominalis, the ashy grey lady beetle, is often a resident of walnut orchards, and during late spring *Hippodamia* may invade walnut trees. It appears that *Olla* remains in orchards having at least three *Chromaphid juglandicola* (Kalto) per leaflet present while four aphids per leaflet are necessary to retain *Hippodamia* (Sluss, 1967).

Adalia bipunctata, twospotted lady beetle, is generally a coastal species in California and becomes active mainly in the summer feeding on various aphids infesting trees and shrubs. No predator-prey ratios have been determined for this species in North America; however, in France releases of five adults per apple tree at 15-day intervals controlled the aphid *Dysaphis plantaginea* (Passerini) (Remaudiere *et al.,* 1973), and in Poland *A. bipunctata* adults are able to suppress the apple aphid at a ratio of 1 up to 250 aphid nymphs and the lady beetle larvae at a ratio of 1 predator to 60 nymphs (Niemczyk *et al.,* 1974).

Syrphidae. The adults of the common syrphid species (fig. 5) usually oviposit in higher aphid populations than coccinellids. If coccinellids are not abundant, the syrphid larvae (fig. 6) usually become the dominant predator. In some springs syrphid larvae are important predators of the green peach aphid, *Myzus persicae* Sulzer, on peach trees in Washington State (Tamaki, 1973). In the autumn, syrphid larvae are the only predators that effectively suppress aphids on peach trees as they can work under cooler conditions than most other aphid predators. Tamaki *et al.* (1967) found that during this season 30 to 40 aphids per leaf provided maximum stimulus for syrphid egg deposition, which accounted for about 26 syrphid larva per 100 leaves. The syrphid adults require pollen from wild flowers or weeds in order to produce eggs. Thus the adults must come from habitats other than peach trees.

Fig. 5. Syrphid adult.

Fig. 6. Syrphid larva.

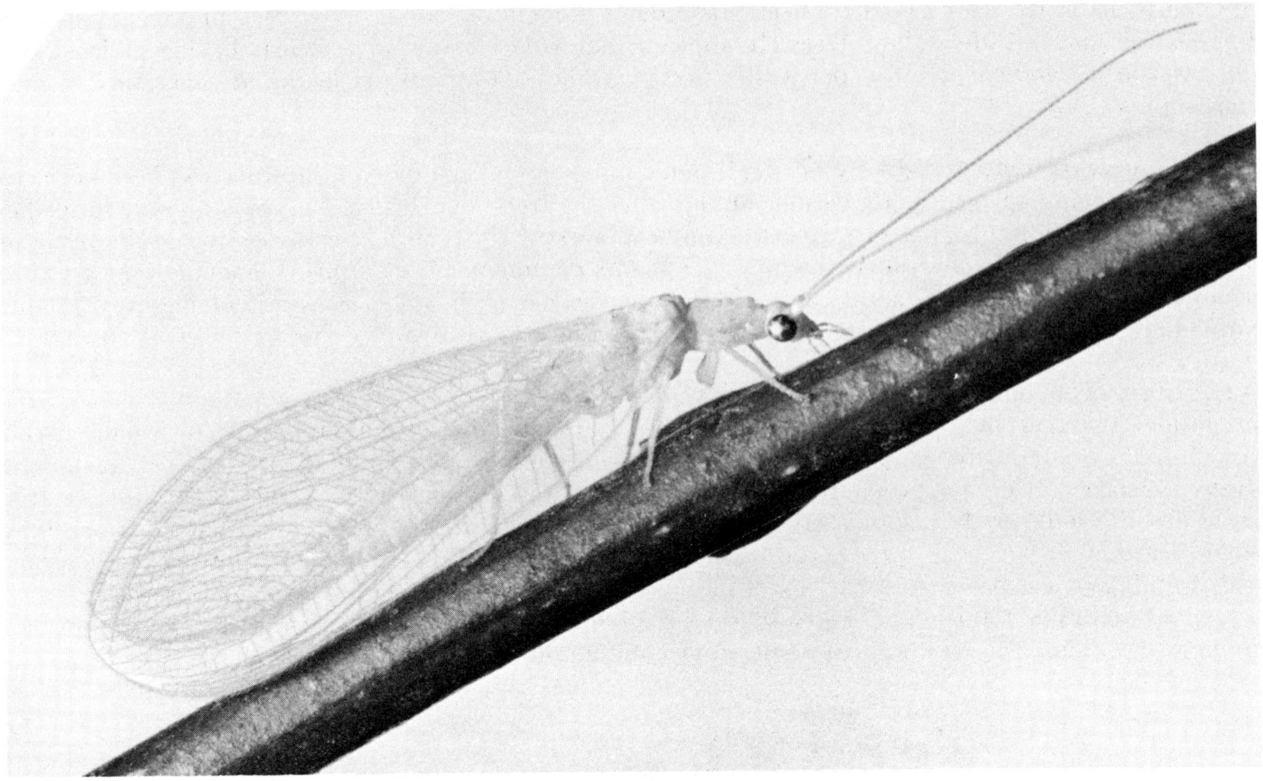

Fig. 7. *Chrysopa carnea* Stephens adult.

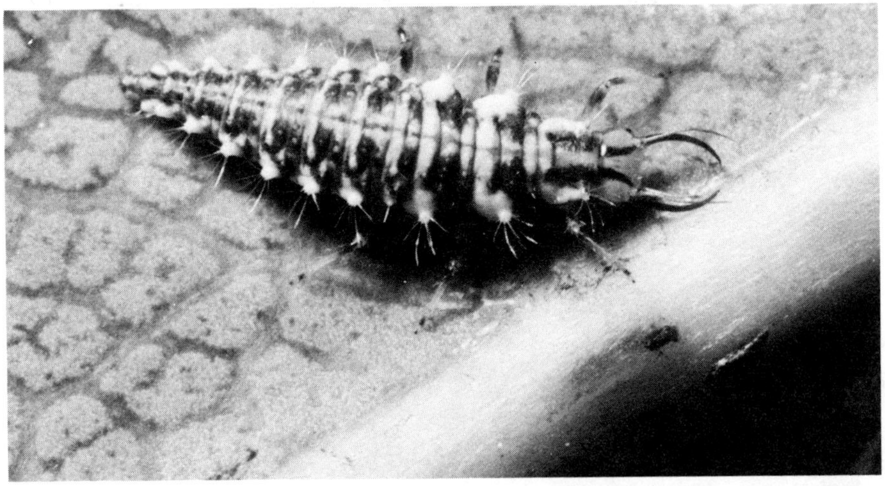

Fig. 8. *Chrysopa carnea* Stephens larva.

Chrysopidae. *Chrysopa* (fig. 7) are either predaceous or non-predaceous in the adult stage, depending upon the species. The latter feed on honeydews, pollens, or nectars. The larvae (fig. 8) of all *Chrysopa* are predaceous. *Chrysopa nigricornis* Burmeister, a common and mostly arboreal species in the United States, is predaceous in the adult stage and must find and eat enough prey to produce eggs. The clusters of eggs are deposited in the immediate feeding area of the adult. *Chrysopa carnea* Stephens, a common holarctic species, is essentially a honeydew feeder. It is attracted to honeydew on which it feeds to produce eggs, and its eggs are singly deposited in the general area of their feeding. Therefore, the aphid or homopterous populations reach rather high levels before *C. carnea* can produce and deposit eggs compared to prey population levels that induce oviposition by coccinellids.

The few *Chrysopa*-prey ratios that have been formulated have come mainly from correlating the numbers of *Chrysopa* eggs or larvae that have been released or increased to control certain prey. A minimum of about 50,000 *C. carnea* eggs per acre released three times controlled tobacco budworm on cotton in Texas (Ridgway and Jones, 1969). With three cotton plants per foot, there are about 39,200 plants per acre. Thus, an average of at least 1.3 eggs per plant resulted in enough *Chrysopa* larvae from three such releases to control *Heliothis virescens* (Fabricius). A ratio of two to three eggs per cotton plant in late August and early September reduced *Heliothis zea* (Boddie) eggs and larvae on cotton in California (Hagen *et al.,* 1971). An average of 1.5 *C. carnea* eggs per bell pepper plant in July and August allowed the green peach aphid to attain 50 aphids per leaf while 3 *C. carnea* eggs per plant in late July allowed an average peak of 25 aphids per leaf (Hagen and Hale, 1974).

Experiments carried out in apple orchards by Niemczyk *et al.* (1974) indicated that 3rd-instar larvae of *C. carnea* are capable of eradicating a population of *Aphis pomi* DeGeer when these occur at a ratio of 1:70 aphid nymphs but not at 1:150 aphid nymphs; one 3rd-instar *C. carnea* larva was able to control the young produced by 24 adult rosy apple aphids, but one larva could not stop the aphid population increase when the ratio was 1:30 adult aphids under caged conditions.

Miscellaneous predators of aphids. The heteropterous bugs that at times feed on aphids in alfalfa are *Nabis, Geocoris* and *Orius.* These general or polyphagous predators apparently do not have the ability to stop increasing aphid populations in spring or fall, but in summer (when aphids are generally scarce) the occasional aphid they may destroy may delay an aphid population resurgence. These general predators like *Chrysopa* larvae will feed upon lepidopterous eggs and larvae, thrips, spider mites, white flies, psyllids and, except for *Chrysopa,* these bugs will feed on lygus eggs and nymphs (Leigh and Gonzales, 1976). If these general predators are reduced by pesticides, particularly in cotton fields, various lepidopterous pests may increase rapidly (Ehler and van den Bosch, 1974; van den Bosch *et al.,* 1971).

Combined effect of predators of aphids. In aphid infestations there usually are mixtures of species and stages of predators feeding on aphids. Therefore, all aphid predators counted in samples should be summed and rated as to their voracity. Tamaki *et al.* (1974) attempted to formulate a predictive predator model for field use. They divided the aphid predator complex into three classes depending upon their voracity, based on similar feeding capacities. Each class is given a value called "predator power." The polyphagous predators (class 1) e.g., nabids, anthocorids, geocorids are rated with a predator power of 1, the aphidophagous larvae of large coccinellids, syrphids, chrysopids and *Scymnus* adults were rated 4 (class 2) and large aphidophagous adult coccinellids rated 8 (class 3). Total predator power is, therefore, the sum of all predators monitored in class 1, plus 4 times the sum of class 2, plus 8 times the sum of class 3.

The total predator power is multiplied by predator efficacy to give projected reduction of aphids. Predator efficacy is affected by temperature, prey density, seasonal variation and host plant conditions that affect predator power by reducing the population of prey. Predator power multiplied by predator efficacy yielded an estimate of the impact of the predator complex on a population of aphids, both with respect to the number of aphids consumed and the potential growth of the aphid population in the absence of predators. The calculated population curves of aphids on potato plants and sugar beets exposed to various predator population densities using the model were similar to those observed in field plots where Tamaki *et al.* (1974) monitored aphid and predator densities.

Parasites of aphids

The most important parasites of aphids are small ichneumonoid wasps belonging to the family Aphidiidae. They lay an egg in an aphid, and the egg hatches into a small larva which devours the inside of the aphid. When the larva matures, it usually remains within the aphid and pupates. An aphid killed by a parasite is easily recognized because the aphid becomes discolored, rigid and inflated forming a

Fig. 9. Mummified pea aphids parasitized by *Aphidius smithi* Sharma and Subba Rao.

Fig. 10. Adult male of *Aphidius smithi* Sharma and Subba Rao.

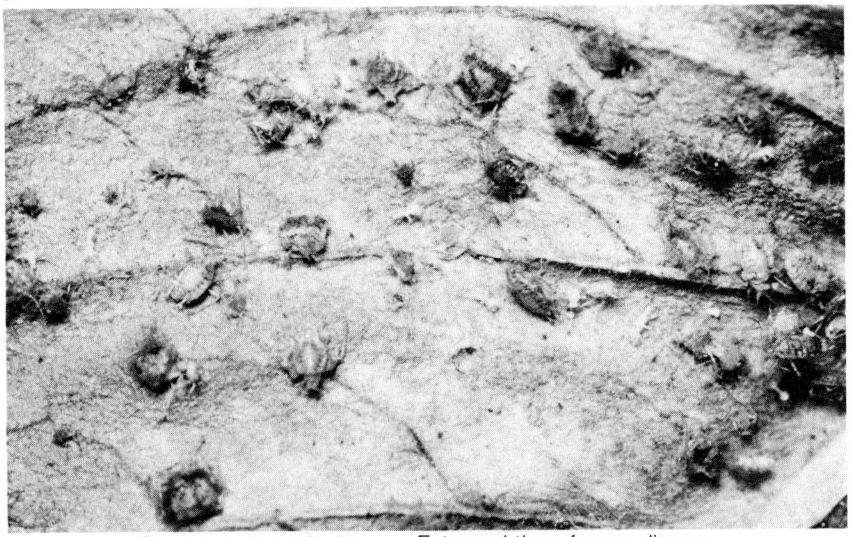

Fig. 11. Aphids killed by an *Entomophthora* fungus disease.

mummified aphid skin (fig. 9). When the parasite leaves the aphid, it cuts a round hole through the aphid mummy and escapes. If mummified aphids are observed it indicates that parasites are at work, but to determine to what extent parasitism is occurring among aphid colonies, it may be necessary to dissect a sample of aphids and look for the legless grubs in the aphid.

Aphidius smithi Sharma and Rao (fig. 10), introduced from India in 1960, was common in California alfalfa fields attacking the pea aphid (PA) (Hagen and Schlinger, 1960; van den Bosch *et al.*, 1967), but it virtually disappeared a few years ago, apparently because of a severe winter, although today (1977) it is once again becoming common. It also attacks the blue alfalfa aphid (BAA). An effective parasite-aphid ratio has not been determined. However, if mummified PA and BAA are observed commonly, the parasites could have a significant impact on the aphid populations.

On walnuts an aphidiid wasp, *Trioxys pallidus* (Haliday), introduced from Iran is controlling the walnut aphid, *Chromaphis juglandicola* (Kaltenbach). Therefore, if mummified walnut aphids are present, the aphid populations should not cause economic damage (Frazer and van den Bosch, 1973).

Pathogenic fungi of aphids. In order for an epizootic (mass-kill) of aphids to be caused by pathogenic fungi, *Entomophthora* spp. (fig. 11), several ecological conditions must coincide: The *Entomophthora* organisms must be present, the aphid population has to be rather high, the temperature should be fairly warm, relative humidity (RH) must be high or rain must occur and, according to Wallace *et al.* (1976), at least for *Entomophthora aphidis* Hoffman a long daylight period (over 14 hours) must prevail for germination of resting spores.

Some specific data involving conditions to set off an epizootic are presented for one aphid species to give some idea of the requirements. MacLeod *et al.* (1966) found that from 4000 to 6000 spores of *Entomophthora aphidis* remained during the winter on a pine needle on which one aphid was killed by the fungus. Where there was a ratio of about 6 of these spore aggregates per 100 aphids, complete mortality occurred in about 12 days. Where there was only one spore aggregate per 100 aphids, complete kill occurred in 17 days. An idea of aphid density necessary for induction of an epizootic is at least 15 aphids per alfalfa stem or 60 to 90 aphids per foot of pine branch to trigger a major epizootic (100 percent mortality). At densities below this threshold, infection occurred in less than 20 percent of the aphids (MacLeod *et al.,* 1966). Temperatures have to be above 20°C for at least some *Entomophthora* spp. to be active.

Epizootics of *Entomophthora* in high aphid populations have been triggered after more than 2 days of rain lasting about 5 hours per day followed by an increase of maximum temperatures above 20°C, or many rainy days followed by relative humidities above 90 percent for more than 10 hours per day (Missonnier *et al.,* 1970).

During cool spring or late fall weather, parasite and predator activity is slowed while the aphids still reproduce actively. Therefore, high aphid populations are generated. Under cool conditions, aphids often exceed economic injury levels before epizootics occur. If there is evidence of *Entomophthora* among the aphids and the relative humidity is not high, irrigation may help to initiate an epizootic. Even though an aphid population may initially exceed an economic injury level, the appearance of a sudden epizootic can account for economic control by preventing sustained feeding by the aphids.

Coleoptera

In alfalfa the only serious coleopterous pests are the *Hypera* weevils. A parasite-host ratio has been determined for *Hypera postica* (Gyllenhal). There is no practical way of distinguishing the alfalfa weevil, *H. postica,* from the Egyptian alfalfa weevil, *H. brunneipennis* (Boheman), in either the adult or larval stages.

The ichneumonid, *Bathyplectes curculionis* (Thomson) is an important parasite of *H. postica.* It totally controls the pest in coastal valleys of central California (Michelbacher, 1940) and is important in Utah (Davis, 1976). Even though it attacks the Egyptian alfalfa weevil and the eastern strain of *H. postica,* it is not as effective as on *H. postica* of central coastal California. An intensive biological control program is being directed at the Egyptian weevil and the alfalfa weevil in the eastern U.S.A. A number of parasites are being introduced against these weevils.

Davis (1974) found that one adult *B. curculionis* per sweep per 25 weevil larvae in Utah produces about 60 percent parasitism, one adult:45 larvae gives 50 percent parasitism and 1 adult:100 larvae accounts for about 35 percent parasitism.

The impact of *B. curculionis* on *H. postica* population density is not evident until adults emerge instead of weevils. Unlike *Apanteles medicaginis* Muesebeck on the alfalfa caterpillar, parasitized weevil larvae feed somewhat less extensively than unparasitized larvae; however, they cannot be distinguished from unparasitized larvae without dissection (Duodu and Davis, 1974). Therefore, in spite of a large number of weevil larvae that may be parasitized it still may be necessary to control the weevil larvae when the larval count reaches an average of 20 per sweep; after larvae begin to appear, fields should be checked at 2-to 4-day intervals. On short alfalfa early in the season, or on stubble following cutting which cannot be checked with a sweep net, treatment is indicated when growth is retarded because of weevil feeding (Bowen *et al.,* 1976).

In Utah, three basic procedures are recommended to cope with *H. postica* (Davis, 1976): (1) evaluate the weevil populations 2 to 3 times during May, (2) use as many cultural practices as possible to reduce weevil numbers, and (3) when chemical treatments are necessary, time the applications and select insecticides to avoid serious damage to parasites, predators and pollinators.

In Utah, Davis (1976) outlines 4 choices of alfalfa weevil control: (1) spray early in the season on a routine basis. This option should be used in areas having a history of severe damage every year plus damage that comes well before cutting time. In much of Utah, however, the major feeding damage is near the time of harvest, and is often disrupted by cutting; also, weevil numbers vary greatly from year to year. This opens up the three options which must be determined on an individual field basis. About 3 weeks before the estimated harvest date, a critical decision must be made. (2) If weevil damage is apparent on 25 percent of the terminals, or there is at least one large larva per alfalfa terminal, a chemical treatment should be made before cutting. About a week before the cutting date, the field should be re-examined. (3) If damage is pronounced on 50 percent of the terminals, cut right away. If there is only a potential problem based on numbers of younger larvae, wait until the normal cutting time, then spray the stubble soon after hay is removed. (4) During this pre-harvest evaluation, if there is no indication of potential injury, forget chemical control entirely.

In areas where the multivoltine eulophid endoparasite, *Tetrastichus incertus* (Ratzeburg), is established on *Hypera* larvae, a ratio of one *T. incertus* adult to eight weevil larvae is necessary to achieve 50 percent parasitism (Horn, 1971). In competition with *B. curculionis,* parasitism by *T. incertus* decreased.

Lepidoptera

In alfalfa there are several potential lepidopterous pests, such as the alfalfa caterpillar, the beet armyworm, yellow-striped armyworm and webworms. However, the only known parasite-host ratio which is practial is one involving the alfalfa caterpillar.

Caterpillars of the alfalfa butterfly, *Colias eurytheme* Boisduval, can sometimes cause serious defoliation of alfalfa in California's Central Valley in summer. This pest would be far more serious if it were not for the parasite *Apanteles medicaginis* Muesebeck, which usually keeps the number of caterpillars below the economic injury level in most fields (Michelbacher and Smith, 1943; Stern and Bowen, 1963). This solitary endoparasite kills the host in early instars before they reach ½ inch (15 mm) in length (fig. 12) so the *Colias* larvae do not have an opportunity to feed extensively before they die. Therefore, by determining how many caterpillars are parasitized early in an alfalfa-cutting period, after the main butterfly oviposition has occurred (alfalfa over 6 inches [15 cm] high), it is possible to predict how many large caterpillars will be present 1 to 2 weeks later before any significant damage could occur. The economic injury level is reached when an average of ten large larvae each over ½ inch long is taken per sweep in a good alfalfa stand (at least 40 stems per square foot). A parasitized caterpillar is lighter than normal in color, and somewhat shiny and swollen toward the rear. By grasping the caterpillar at each end of the swelling and pulling it apart, the small, shiny, whitish parasite larva will pop out. The parasitized caterpillars are subtracted from total number taken by sweeping. If an average of 10 or more caterpillars are not parasitized per sweep, some control measure is advised.

Fig. 12. Alfalfa butterfly caterpillar, *Colias eurytheme,* (lower) showing emergence hole made by the parasitic larva *Apanteles medicaginis* (upper) which has eaten out the caterpillar and is beginning to spin a cocoon.

In northern and eastern states where *Colias philodice* Godart predominates, the parasite *Apanteles flaviconchae* Riley is the main natural enemy, but since this parasite is gregarious (more than one parasite per host) and emerges from larger caterpillars than *A. medicaginis,* these parasitized caterpillars will cause some damage before they are killed.

Tetranychidae Predator-Prey Ratios

Acarina: Phytoseiidae

The phytoseiid mites are important predators of spider mites. They can be distinguished from their prey by their more rapid movements. In general they are shiny, unspotted, oval to pear-shaped and their setae are less conspicuous than those of the phytophagous spider mites. The common phytoseiid mites have slightly oval eggs compared to perfectly sperical eggs of the tetranychid mites. Some practical predator-prey ratios have been formulated for spider mite pests of apples, pears, avocado, citrus and grapes. Predator-spider mite ratios have also been proposed for a mirid and a coccinellid.

Typhlodromus occidentalis Nesbitt. Data of Croft and Nelson (1972) from apple trees in southern California show that if ratios of this species to *Tetranychus mcdanieli* McGregor did not fall below about 1:20 the predator reduced populations of *T. mcdanieli* before the latter reached a density of 20 active stages per leaf. This was considered economic control. It was possible to make predictions as to the outcome of the predator-prey interaction, and to make decisions on whether chemical control was necessary. Ratios of *Tetranychus mcdanieli* to *Typhlodromus occidentalis* are useful only when actual numbers of *T. mcdanieli* exceed five per leaf (Hoyt, unpublished). Dispersion and relative rates of population development can alter the ratios tremendously when actual numbers are low.

In Washington a temperature-driven simulation model of *T. mcdanieli* populations has been developed (Logan, 1977). This model includes a predation component. The required inputs for a simulation include the numbers of predators *(T. occidentalis)* and prey. The output is a complete simulation of the populations of the two species for the remainder of the interaction (usually the season) based on chosen temperature projections. The simulations have very closely approximated the actual population events, with some differences in predicted peak populations.

T. occidentalis is also an important predator of *Tetranychus pacificus* McGregor on grapes. Ratios from 1:30 to 1:40 are favorable for the control of Pacific mites on grapes in central California (Flaherty and Lynn, unpublished; AliNiazee *et al.*, 1974).

Amblyseius hibisci (Chant). Increases in numbers of this species are sometimes correlated with the presence of pollen and reproduction on this food is more rapid than on spider mites (McMurtry and Scriven, 1964). Kennett and Flaherty (1974) suggested that a ratio of one *A. hibisci* to six *Panonychus citri* (McGregor) was required in the early spring on citrus in the San Joaquin Valley to prevent *P. citri* population from increasing. In this area the presence of wind-blown pollens greatly influences the population of *A. hibisci*. In southern California a ratio of one *A. hibisci* to about seven *P. citri* (all active stages) apparently was sufficient for the predator to suppress the *P. citri* population.

Amblyseius limonicus Garman and McGregor. In a greenhouse experiment, this predator controlled *Oligonychus punicae* (Hirst) at about the same level when the starting predator:prey ratios were 1:7 or 1:23. During the interaction, the ratios became as low as 1:60.

Phytoseiulus persimilis Athias-Henriot. This predator controlled *Tetranychus urticae* (Koch) on clover in a greenhouse experiment when the starting predator:prey ratio was as low as 1:64; quicker control at lower levels occurred when the initial ratio was 1:32 (Mori and Moriyama, 1970). Evidently phytoseiid species vary considerably in regard to their ability to "catch up" numerically and suppress a tetranychid mite population. Inherent characteristics which could be responsible for these differences include reproductive capacity, prey consumption capacities, mobility, and dependence on the prey.

Hemiptera

Blephariodopterus angulatus (Fall.) (Miridae). This univoltine species is common in apple trees in England, and it was estimated that a ratio of at least 1 predator to 800 *Panonychus ulmi* (Koch) was sufficient to prevent increase of this mite. If the ratio fell below this level, *P. ulmi* populations increased (Muir, 1965).

Coleoptera

Stethorus picipes Casey. This species is considered the key predator in the control of *Oligonychus punicae* (Hirst) on avocado (McMurtry and Johnson, 1966). Judging by the numbers of *S. picipes* counted in a 5-minute search on each sample tree, and the number of adult female *O. punicae* counted in a 10-leaf sample per tree, short-term predictions could be made on whether *O. punicae* would be suppressed before considerable leaf damage occurred. If the predator index was about 15 to 20 per 5-minute count when the mite density approached 10 adult females per leaf, or was 30 to 40 per 5-minute count at a mite density of 20 per leaf, control of the mites occurred at a tolerable level.

Hull *et al.* (1976) showed that the distribution of *Stethorus punctatum* (LeConte) in an apple orchard and within an apple tree is closely related to the distribution of European red mite, *Panonychus ulmi* (Koch). A prey population of two to three mobile mites per leaf was sufficient to enable the predator population to reproduce and increase in size.

Neuroptera

Although there apparently have been no *Chrysopa* larva/spider mite ratios determined, there are data available on the number of *Chrysopa carnea* first and second instar larvae per so many leaves necessary to control *P. ulmi* on apple trees (Niemczyk *et al.,* 1973). If at least one chrysopid larva occurs per 10 to 25 leaves, the predator is able to limit and maintain the mite populations below economic injury levels. One *Chrysopa* larva per 50 leaves is borderline and allows an average of 5 mites per leaf, and with 1 per 75 leaves the mites are not controlled. In order to obtain oviposition by *Chrysopa carnea,* however, some homopterous honeydew has to be available in the orchard or cover crop.

Acknowledgment

We wish to thank F. E. Skinner of the Division of Biological Control, Department of Entomological Sciences, University of California, Berkeley, for taking and providing all of the photographs of insect predators presented in this paper.

Literature Cited

ALINIAZEE, M. T., E. M. STAFFORD and H. KIDO

1974. Management of grape pests in central California vineyards: toxicity of some commonly used chemicals to *Tetranychus pacificus* and its predator, *Metaseiulus occidentalis.* J. Econ. Entomol. 67: 543-47.

BOWEN, W. R., V. E. BURTON, W. R. COTHRAN, L. A. FALCON, K. S. HAGEN, I. M. HALL, H. T. REYNOLDS, V. M. STERN, C. G. SUMMERS, J. E. SWIFT, N. C. TOSCANO, and R. van den BOSCH.

1976. Pest control program for alfalfa hay. Div. Agric. Sci. Univ. Calif. Leaflet 2763. 20 pp.

COLLYER, E.

1958. Some insectary experiments with predacious mites to determine their effect on the development of *Metatetranychus ulmi* (Koch) populations. Entomologia exp. appl. 1: 138-46.

COLLYER, E. and A. H. M. KIRBY.

1955. Some factors affecting the balance of phytophagous and predacious mites on apple in south-east England. J. Hort. Sci. 30: 97-108.

CROFT, B. A. and E. E. NELSON.

1972. An index to predict efficient interactions of *Typhlodromus occidentalis* in control of *Tetranychus mcdanieli* in southern California apple trees. J. Econ. Entomol. 65: 310-12.

DAVIS, D. W.

1974. Parasite-prey ratios among alfalfa weevil larvae of northern Utah. Environ. Entomol. 3: 1031-32.

DAVIS, D. W. (editor)

1976. Insects and nematodes associated with alfalfa in Utah. Utah Agric. Exp. Sta. Bul. 469. 59 pp.

DUODU, Y. A. and D. W. DAVIS.

1974. Effects of *Bathyplectes curculionis* (Thomson) on the development, morphological appearance and activity of alfalfa weevil larvae. Environ. Entomol. 3: 396-98.

EHLER, L. E. and R. van den BOSCH.

1974. An analysis of the natural biological control of *Trichoplusia ni* (Lepidoptera: Noctuidae) on cotton in California. Can. Entomol. 106: 1067-73.

FRAZER, B. D. and R. van den BOSCH.

1973. Biological control of the walnut aphid in California: the interrelationship of the aphid and its parasite. Environ. Entomol. 2: 561-68.

HAGEN, K. S.

1962. Biology and ecology of predaceous Coccinellidae. Ann. Rev. Entomol. 7: 289-326.

1974. The significance of predaceous Coccinellidae in biological and integrated control of insects. Entomophaga Memo. H. S. 7: 25-44.

1976. Role of nutrition in insect management. Proc. Tall Timbers Conf. Ecol. Anim. Control by Habitat Mangmt. 6: 221-61.

HAGEN, K. S. and R. HALE.

1974. Increasing natural enemies through use of supplementary feeding and non-target prey. *In* F. G. Maxwell and R. A. Harris (eds.), Proceedings of the Summer Institute of Biological Control of Plant Insects and Diseases. Univ. Press, Jackson, Miss. pp. 170-81.

HAGEN, K. S., E. F. SAWALL, JR. and R. L. TASSAN.

1971. The use of food sprays to increase effectiveness of entomophagous insects. Proc. Tall Timbers Conf. Ecol. Anim. Control Habitat Mangmt. 2: 59-81.

HAGEN, K. S. and E. I. SCHLINGER

1960. Imported Indian parasite of pea aphid established in California. Calif. Agric. 14: 5-6.

HAGEN, K. S. and R. R. SLUSS.

1966. Quantity of aphids required for reproduction by *Hippodamia* spp. in the laboratory. *In* I. Hodek (ed.), Ecology of aphidophagous insects. Prague: Academia. pp. 47-59.

HAGEN, K. S. and R. van den BOSCH.

1968. Impact of pathogens, parasites and predators in aphids. Ann. Rev. Entomol. 13: 325-84.

HORN, D. J.
1971. The relationship between a parasite, *Tetrastichus incertus* (Hymenoptera : Eulophidae), and its hosts the alfalfa weevil, *Hypera postica* (Coleoptera: Curculionidae) in New York. Can Entomol. 103: 83-94.
HULL, L. A., D. ASQUITH, and P. D. MOWERY.
1976. Distribution of *Stethorus punctum* in relation to densities of the European red mite. Envir. Entomol. 5: 337-42.
KENNETT, C. E. and D. L. FLAHERTY.
1974. Pest management of citrus and red mite on citrus in the southern San Joaquin Valley. San Joaquin Valley Agric. Res. and Extension Center, Parlier, Calif. 30 pp.
LEIGH, T. F. and D. GONZALES.
1976. Field cage evaluation of predators for control of *Lygus hesperus* Knight on cotton. Environ. Entomol. 5: 948-52.
LOGAN, J. A.
1977. Population model of the association of *Tetranychus mcdanieli* (Acarina: Tetranychidae) with *Metaseiulus occidentalis* (Acarina: Phytoseiidae) in the apple ecosystem. Ph.D. Thesis, Wash. State Univ.
MacLEOD, D. M., J. W. M. CAMERON and R. S. SUPER
1966. The influence of environmental conditions on epizootics caused by entomogenous fungi. Rev. Roumaine Biol. Botan. 2: 125-34.
McMURTRY, J. A. and H. G. JOHNSON
1966. An ecological study of spider mite *Oligonychus punicae* (Hirst) and its natural enemies. Hilgardia 37: 363-402.
McMURTRY, J. A. and G. T. Scriven
1964. Studies on the feeding, reproduction, and development of *Amblyseius hibisci* (Acarina: Phytoseiidae) on various food substances. Ann. Entomol. Soc. Amer. 57: 649-55.
MICHELBACHER, A. E.
1940. Effect of *Bathyplectes curculionis* on the alfalfa-weevil population in lowland middle California. Hilgarida 13: 81-.
MICHELBACHER, A. E. and R. F. SMITH
1943. Some natural factors limiting the abundance of the alfalfa butterfly. Hilgardia 15: 369-97.
MISSONNIER, J., Y. ROBERT and G. THORZON
1970. Circonstances epidemiologiques semblant favorisor le development des mycoses a Entomopthorales chez trois aphides. *Aphis fabae* Scop., *Capitophorus horni* Borner et *Myzus persicae* (Sulz.). Entomophaga 15: 169-90.
MORI, H. and S. MORIYAMA
1970. Biological control of *Tetranychus urticae* Koch on white clover by *Phytoseiulus persimilis* Athias-Henriot (Acarina: Tetranychidae, Phytoseiidae). Mushi 43: 29-38.
MUIR, R. C.
1965. The effect of sprays on the fauna of apple trees. III. The influence of certain fungicides and peta-fall insecticides on the fruit tree red spider mite, *Panonychus ulmi*, (Koch) and its predators. Rep. E. Malling Res. Sta. for 1964: 162-70.
NEUENSCHWANDER, P., K. S. HAGEN and R. F. SMITH.
1975. Predation on aphids in California's alfalfa fields. Hilgardia 43: 53-77.
NIEMCZYK, E., M. MISZCZAK, R. OLSZAK, S. SLEDZIONE and G. BAKOWSKI.
1973. The effectivenss of some predaceous insects in the control of phytophagous mites and aphids on apple trees. Ann. Report (1973) Research Instit. Pomology, Skierniewice, Poland. 49 pp.
NIEMCZYK, E., R. OLSZAK, M. MISZCZAK and G. BAKOWSKI.
1974. Effectiveness of some predaceous insects in the control of phytophagus mites and aphids on apple trees. Ann. Rep. 1974. Res. Instit. of Pomology, Poland. 84 pp.
REMAUDIERE, G., G. IPERTI, F. LECLANT, J. P. LYON and M. F. MICHEL.
1973. Biologie et ecologie des aphids et de leurs enemis naturels. Application a la lutte integee en vorgers. Entomophaga, Mem. H. S. 6. 35 pp.
RIDGWAY, R. L. and S. L. JONES.
1969. Inundative release of *Chrysopa carnea* for control of *Heliothis* on cotton. J. Econ. Entomol. 62: 177-80.
SLUSS, R. R.
1967. Population dynamics of walnut aphid *Chromaphis juglandicola* (Kalt.) in northern California. Ecology 48: 41-58.
SMITH, R. F. and K. S. HAGEN.
1966. Natural regulation of alfalfa aphids in California. *In* I. Hodek (ed.), Ecology of aphidophagous insects. Prague: Academia. pp. 297-315.
STERN, V. M. and W. BOWEN.
1963. Ecological studies of *Trichogramma semifumatum,* with notes on *Apanteles medicaginis,* and their suppression of *Colias eurytheme* in southern California. Ann. Entomol. Soc. Amer. 56: 358-72.
TAMAKI, G.
1973. Spring populations of the green peach aphid on peach trees and the role of natural enemies in their control. Environ. Entomol. 2: 186-91.
TAMAKI, G., B. J. LANDIS and R. E. WEEKS.
1967. Autumn populations of green peach aphid on peach trees and role of syrphid flies in their control. J. Econ. Entomol. 60: 433-36.
TAMAKI, G., J. V. McGUIRE and J. E. TURNER.
1974. Predator power and efficacy: a model to evaluate their impact. Environ. Entomol. 3: 625-30.
van EMDEN, H. F. (editor).
1972. Aphid Technology. Academic Press, London, N. Y., 344 pp.
van den BOSCH, R., C. F. LAGACE and V. M. STERN.
1967. The interrelationships of the aphid *Acyrthosiphum pisum* and its parasite *Aphidius smithi* in a stable environment. Ecology 48: 993-1000.
van den BOSCH, R., T. F. LEIGH, L. A. FALCON, V. M. STERN, D. GONZALEZ and K. S. HAGEN.
1971. The developing program of integrated control of cotton pests in California. *In* C. B. Huffaker (ed.) Biological Control, Plenum Press, N. Y., London.
WALLACE, D. R., D. M. MacLEOD, C. R. SULLIVAN, D. TYRRELL and A. J. DeLYZER.
1976. Induction of resting spore germination in *Entomophthora aphidis* by long-day light conditions. Can. J. Bot. 54: 1410-18.

Section II
PEST MANAGEMENT METHODS

Chapter 5.
INTRODUCTION OF NEW SPECIES AND BIOTYPES

L. A. Andres and J. A. McMurtry

Introduction

Natural enemies play a key role in the pest management program. Where natural enemies of a particular insect pest or weed are absent or ineffective, it has been sound practice to search out and import new parasites and predators from other areas of the pest's range. Historically, this has been the major approach in biological control. Two recent examples can be cited. Three parasitic wasps, introduced from the Middle East, substantially aid in the control of the naturalized spotted alfalfa aphid, (*Therioaphis maculata* [Buck]) in the western United States. Also, three weed-feeding insects introduced from Argentina into southeastern United States now provide excellent control of the emersed aquatic alligatorweed, *Alternanthera philoxeroides* (Mart.) Griseb., a plant native to South America.

In essence, the procedure involved is one of searching out suitable natural enemies and shipping them to the problem area. However simple this might seem, the successful outcome of each project not only depends on finding potentially effective parasites and predators, but finding ones that will establish and increase to control levels under the environmental conditions particular to the problem site. To accomplish this requires detailed knowledge of the pest (e.g., precise identification, biology, habitat, etc.) and of its natural enemies and the factors regulating their interaction.

Normally, several steps are involved in undertaking a biological control project: (1) The problem must be assessed for amenability to biological control. The characteristics of the problem are examined and the likelihood and effort needed to achieve success determined. (2) Literature and background studies on the pest and its enemies are made and the most promising foreign areas selected for search and study. (3) Collecting, studying and introducing the natural enemies into quarantine in the United States. (4) Finalizing clearance tests, releasing and establishing and conducting follow-up studies.

Even though an exotic natural enemy may become naturalized in a new area, pest control may still be less than satisfactory for various reasons. In some instances control may be improved through supplemental or periodic releases of the natural enemies, by using supplementary foods to encourage immigration and build-up of natural enemies in specific problem areas, by altering cultural practices that may otherwise act to the detriment of natural enemies, and/or the judicious use of selective pesticides to minimize harm to the parasites. These aspects of the pest management program are discussed in some detail in other papers in this publication.

Amenability of the Pest Problem to Biological Control

Not all pests are equally suited for control by introduced natural enemies. The cause of each problem must be assessed. Native insect and weed pests often have existing complements of natural enemies which are rendered ineffective due to cultural practices or other reasons. Whether pest problems of this nature can be solved by introducing additional natural enemies is often unclear, and in many instances unlikely. On the other hand, the probabilities are generally good for controlling introduced pests by introducing their parasites and predators.

Seemingly, biological control has been more successful against insect pests of perennial plants, especially those pests that feed externally on the plant and are more or less sessile in habit. Likewise there have been more successful attempts in controlling perennial weeds in areas of low disturbance (e.g., range and pasture areas) than weeds among crop plants.

Biological control does not eradicate the target pest. In fact, the continual presence and effectiveness of the natural enemy depends on the low level presence of the host. Pest insects attacking the nonmarketable portion of the plant (i.e., the foliage on a citrus or apple tree) or "indirect" pests, are often thought to be better suited to biological control. However, the successful introduction of parasites from Iran and Pakistan against the olive scale, a "direct" pest on the olive fruit, suggests that even this type of pest is amenable to biological control. Cullage of olives due to scale on the fruit was reduced from levels sometimes exceeding 50 percent to 0.5 percent or lower by the parasites (DeBach *et al.,* 1971).

Host-specific insects have been quite useful in controlling weedy plants in many areas. However, the more similar the weed is to related plants of economic importance, the more difficult it is to find insects that will limit their attack to the weed and not harm the plants of value. For example, artichokes are closely related to thistles, and are susceptible to attack by several insects which could otherwise be used in biological control of thistles. Conflicting interests over whether a weed is really a weed may also influence its amenability to biological control. For example, yellow starthistle is valued by beekeepers in California, but it reduces productivity of pastures and grain fields. The losses and benefits of each weed are carefully considered and conflicting interests resolved before biological control is initiated.

Thus the amenability of a particular pest to biological control is neither clear-cut nor always predictable. As is often the case, the degree of success achieved through biological control is in direct proportion to the research effort expended. Success frequently becomes a question of resource limitations and the effort we can direct towards the problem.

Background Studies

Before initiating a program for importing exotic natural enemies, a detailed assessment should be made to determine what natural enemies may already be present, or might be present but suppressed by the use of pesticides or other cultural practices. Sprayed as well as unsprayed areas should be examined.

Precise identification of the pest, its distribution throughout the world and the best sites for searching out natural enemies should be determined. If the native home of the pest is known, this represents the logical place to search as it is here that a full complement of natural enemies has most likely evolved, and it is here where the probability for finding an effective natural enemy or complex of enemies is greatest.

Because climate plays an important role in the distribution and abundance of the pest insect, as well as its natural enemies, the search for parasites and predators should include foreign areas climatically similar to those areas where releases are planned. The importance of collecting natural enemies climatically adapted to the planned release area was illustrated during a study on the walnut aphid (*Chromaphis juglandicola* [Kalt.]). An aphid parasite, *Trioxys pallidus* (Haliday), was introduced from France in 1959 and established in southern California where it provided good control along the coast and in nearby valleys (van den Bosch *et al.,* 1962). Failure to establish this parasite in central California led to further exploration and the discovery of a "strain" or biotype of this same parasite in an area of Iran climatically similar to that of California. The Iranian strain of *T. pallidus* was imported in 1968 and readily established in central California, where it is now providing excellent control (van den Bosch *et al.,* 1970).

If we compare the climate of central California and much of the western United States with the climate of other areas of the world, we see that it is similar to that of the eastern Mediterranean (Turkey, Lebanon, Israel) and the Middle East (Iraq, Iran, Afghanistan, Pakistan and central Asia). The dissimilarity of climate between Europe and the western United States may in part account for the limited success and even failure of many parasite introductions made from Europe to the West.

Introduction Protocol

A detailed coverage of this subject is beyond the scope of this book, so for further information the reader is referred to texts by DeBach (1964) or van den Bosch and Messenger (1973). The greater the variety of areas included in the foreign exploration program the greater the chance of finding natural enemies adapted to the particular habitat. Here again success is directly proportional to the time and effort expended. By limiting foreign exploration, key parasite species may be overlooked due to seasonal scarcity and absence due to pesticide uses. Foreign cooperators are especially valuable in locating the desired host plants and pest infestations free from frequent pesticide applications, as well as assisting the collector in overcoming communication and bureaucratic problems. Productive collecting sites often include parks, botanical gardens, backyard plants and roadsides (e.g., for weed enemies), as well as crop areas.

The collector must be familiar with the pest species as well as with the general groups of natural enemies likely to be found. Parasites are usually obtained in the form of larvae or pupae inside or on the host pest. Predators on the other hand can usually be collected as adults or immature stages directly from the plant. The live beneficial organisms are packaged and shipped to this country as soon as possible after collection. Some form of food (such as honey) is added to the container in order to prolong the lives of adults. All possible precautions are taken in packaging to preclude escape of any organisms. Federal permits are required to import beneficial species and their associated hosts, and the package must be opened in one of the several Federal or State quarantine facilities. The parasites are cultured in quarantine and when a reproducing colony is established, and determined to be free of any potentially harmful organisms (diseases, pest species, hyperparasites), permission for its release is granted by state and federal authorities. The beneficial species may be released in the field directly from quarantine or bred in the insectary to increase its numbers before field releases are made.

In the case of insects intended for biological control of weeds, importation is prohibited without prior data on the host range of the candidate species. Therefore, prior to shipment to the United States, extensive feeding tests are conducted overseas to assure that plants of value will not be attacked. Additional feeding studies may be required in quarantine prior to release.

Colonization, Establishment and Utilization

In attempting to establish the new natural enemies, releases are made at various times in a variety of sites to maximize the possibility that conditions necessary for establishment are met. If the range of the pest extends through various climatic areas, colonizations are made in all areas and repeated at different times of the year when suitable stages of the host species are available.

Establishment in the new environment may occur quickly with some species, only after considerable time and massive releases with others, and not at all with still other species. If the introduced species shows promise, it may be desirable to hasten its spread and establishment to other areas. Growers, county agents and other concerned persons frequently assist in large-scale collections from established sites for dissemination in other pest infestations.

Some Recent Introductions of Beneficial Insects in the Western United States

The University of California (Riverside and Albany) and the U.S. Department of Agriculture (Albany) operate quarantine facilities for the study and culture of exotic natural enemies for eventual release against insect and weed pests.

The W-84 project has been involved with natural enemy introductions and colonizations against the following pests:

Egyptian alfalfa weevil, *Hypera brunneipennis* (Boheman). Natural enemies released include *Bathyplectes curculionis* (Thomson), *B. anurus* (Thomson), *B. stenostigma, Microctonus aethiops* (Nees), *M. aethiopoides,* and *Tetrastichus incertus* (Ratzeburg), at colonization sites in California, Colorado, Idaho, Oregon and Utah. Although recovery attempts to date have been negative, this program is continuing.

Psyllids. Several species of Coccinellidae, including *Harmonia conformis* (Boisduval), *H. dimidiata* Fabr., and *Diomus pumilio* (Weise), have been introduced and released against the pear psylla, *Psylla pyricola* Foerster in California, Utah and Washington and the acacia psyllid, *Psylla uncatoides* (Ferris and Klyver) in Hawaii. *H. conformis* is now established in Hawaii and has reduced the acacia psyllid to non-damaging levels at survey sites (Leeper and Beardsley, 1976).

Spider mites. *Phytoseiulus persimilis* Athias-Henriot, a predaceous phytoseiid mite, is established in Ventura County, California, where it plays a role in pest management of twospotted spider mite, *Tetranychus urticae* (Koch), on strawberries grown as an annual crop. Mite-infested weeds or citrus trees serve as reservoirs from which the predators move to the new crop of strawberries. *P. persimilis* and *Typhlodromus occidentalis* Nesbitt have been introduced to Hawaii and Guam against *Tetranychus cinnabarinus* (Boisduval) in papaya and passion fruit orchards. Establishment of both predator species apparently has been achieved in Hawaii.

Amblyseius stipulatus Athias-Henriot, a phytoseiid mite from the Mediterranean region, has been established in a lemon orchard in coastal southern California since 1971, having replaced the common native species, *Amblyseius hibisci* (Chant) on release trees and some adjacent trees. This predator appears to be slowly spreading within an acre block.

Tansy ragwort, *Senecio jacobaea* L.

The European moth, *Tyria jacobaeae* L. has reduced vigorous infestations of Tansy ragwort to small, non-seeding depauperate stands in some areas of California and Oregon. A recent attempt to improve control has been somewhat successful following the introduction of the flea beetle, *Longitarsus jacobaeae* (Waterhouse) in 1968-70. The beetle is now well established in the original release areas (Ft. Bragg, Mendocino County) and has reduced plant rosette density by 90 percent. The beetles have been distributed within California and Oregon.

Diffuse knapweed, *Centaurea diffusa* Lam.

A seed fly (*Urophora affinis* [Frauenfeld]) reared from *Centaurea maculosa* Lam. collections in Europe has been released in Oregon, Washington, Idaho and Montana against *C. diffusa* and/or *C. maculosa* (1974-75). The fly is established in all states, but its eventual impact on the weed hosts is yet unclear.

Mediterranean sage, *Salvia aethiopis* L.

The rosette-infesting weevil, *Phrydiuchus tau* Warner, recently imported from Yugoslavia, has been successfully established in Oregon and is building up to high population levels. The eventual impact on the host plant is not yet known.

Musk thistle, *Carduus nutans* L.

This thistle is one of the ten most serious weeds of range and pasturelands in the United States. Because of its large size and spiny nature, it is avoided by livestock during grazing. In the early 1970's *Rhinocyllus conicus* Froelich was introduced to the United States from the weed's native European area, and it has become well established in Montana and Virginia where it destroys significant amounts of thistle seed. The weevil is being released in many other thistle-infested states.

Russian thistle, *Salsola iberica* Sennen et Pan

This plant is widespread in the western and midwestern United States where it serves as an alternate host for the beet leafhopper, as well as posing a nuisance along highways, canals and in urban areas. In 1973 a stem-mining moth, *Coleophora parthenica* Meyrick, was introduced from Pakistan to California where it has now become established. Distribution to other states is underway.

Availability of Natural Enemies to Growers

Because the self-perpetuating nature of most beneficial insects makes them of little interest to commercial enterprise for distribution and sale, federal and state researchers and extension specialists make every effort to facilitate distribution to interested farmers on request.

Literature Cited

DEBACH, P. (ed.).
1964. Biological Control of Insect Pests and Weeds. Reinhold Publ. Co., N.Y. 844 pp.

DEBACH, P., D. ROSEN and C. E. KENNETT.
1971. Biological control of coccids by introduced natural enemies. Chapter 7. *In:* Biological Control, C. B. Huffaker (ed.), Plenum Press, N.Y. 511 pp.

LEEPER, J. R. and J. W. BEARDSLEY.
1976. The biological control of *Psylla uncatoides* (Ferris and Klyver) (Homoptera: Psyllidae) in Hawaii. Proc. Hawaii. Entomol. Soc. 22:305-19.

VAN DEN BOSCH, R. and P. S. MESSENGER.
1973. Biological Control. Intext Publishers, Inc., New York, N.Y. 180 pp.

VAN DEN BOSCH, R. E. I. SCHLINGER and K. S. HAGEN.
1962. Initial field observations in California on *Trioxys pallidus* (Haliday), a recently introduced parasite of the walnut aphid. J. Econ. Entomol. 55:857-62.

VAN DEN BOSCH, R., B. D. FRAZER, C. S. DAVIS, P. S. MESSENGER and R. HOM.
1970. *Trioxys pallidus*—an effective new walnut aphid parasite from Iran. Calif. Agric. 24(11):8-10.

Chapter 6.

Supplementary Releases of Natural Enemies

E. R. Oatman and J. W. Beardsley

Introduction

In biological control, a supplementary release of an effective natural enemy (parasite or predator) is used to regulate a pest population when sufficient numbers are not present or when natural enemies are not present early enough to provide adequate control. Such releases have been referred to as periodic colonizations, periodic releases, programmed releases, and strategic releases. Their role in relation to other methods of biological control was discussed by DeBach and Hagen (1964) and Ridgway (1972).

To be successful, supplementary releases require a thorough knowledge of the crop ecosystem, pest (host or prey), parasites or predators and their interactions with their hosts, and the ability to economically produce and release predictable numbers of a given parasite or predator of known quality. Natural enemy production is usually less economical than other methods of pest control, such as pesticide applications, and cost is the principal limitation. Ridgway (1972) and Ridgway *et al.* (1974) discussed these and other characteristics of supplementary releases in respect to their application.

Examples

Experimental or limited practicality

Two of the earliest documented cases of supplementary releases of natural enemies to control a pest were the release of green lacewing (then known as *Chrysopa californica* Coq.) eggs on unsprayed pear trees in California to control a mealybug, *Pseudococcus maritimus* (Ehr.) (Doutt and Hagen, 1949), and the release of both phytophagous (cyclamen mite, *Steneotarsonemus pallidus* [Banks] and predaceous mites (*Typhlodromus reticulatus* Oud. and *T. bellinus* Womer) on strawberry in California to obtain a desirable balance between pest and predator populations (Huffaker and Kennett, 1956).

In California, Oatman *et al.* (1968) obtained effective biological control of the two-spotted mite, *Tetranychus urticae* (Koch) on strawberry, using mass releases of a predaceous mite, *Phytoseiulus persimilis* Athias-Henriot, at a rate equivalent to 320,000 per acre. In subsequent similar experiments (unpublished), effective control was obtained with 250,000 *P. persimilis* per acre. McMurtry *et al.* (1969) demonstrated that mass releases of a coccinellid predator, *Stethorus picipes* Casey, at a rate equivalent to 32,000 per acre, resulted in substantially less damage to avocado leaves by the avocado brown mite, *Oligonychus punicae* (Hirst), and a lowering of pest population. On both strawberry and avocado, the number of predators released, timing of releases, and period of time over which the supplementary releases were made were critical to the success of the experiments.

In Texas, Ridgway and Jones (1969) successfully controlled the bollworm, *Heliothis zea* (Boddie), and the tobacco budworm, *H. virescens* (F.), on cotton in small field plots, using supplementary releases of larvae of the green lacewing, *Chrysopa carnea* Stephens. Lingren (1970) successfully controlled these two pests on cotton with releases of *Trichogramma* sp.

In Missouri, Parker *et al.* (1971) obtained effective control of the imported cabbageworm, *Pieris rapae* (L.), on cabbage by releasing fertile imported cabbageworm adults, an egg parasite, *Trichogramma evanescens* Westwood, and a larval parasite, *Apanteles rubecula* Marshall, early in the growing season.

In California, Oatman and Platner (1971) obtained effective control of the tomato fruitworm, *Heliothis zea* (Boddie), the cabbage looper, *Trichoplusia ni* (Hübner), and hornworms, *Manduca* spp., on processing tomatoes (mechanically harvested varieties) using supplementary releases of an indigenous egg parasite, *Trichogramma pretiosum* Riley, at rates equivalent to 465,000 per acre. In subsequent similar experiments (unpublished), effective control was obtained at 200,000 per acre.

In all of the above cited cases of successful supplementary releases, none has been widely accepted for practical application, with the possible exception of *Chrysopa* and *Trichogramma* releases on cotton and tomato.

Commercial practices

Extensive studies in Europe resulted in the development of an effective release program for control of the twospotted spider mite, greenhouse whitefly, *Trialeurodes vaporariorum* (Westwood), and the cotton aphid, *Aphis gossypii* Glover, on cucumbers grown in greenhouses (Hussey and Bravenboer, 1971). The predaceous mite, *Phytoseiulus persimilis,* is released for control of the twospotted spider mite; and the parasites, *Encarsia formosa* Gahan and *Aphelinus flavipes* Kurdjumov, for control of the greenhouse whitefly and cotton aphid, respectively. The procedure used is briefly as follows: at specified time intervals after planting, specified numbers of prey and/or host are released on the plants followed by releases of specified numbers of predator and/or parasite at specified intervals. Strict adherence to the program results in increased yields, due partly to better mite control and to the elimination of phytotoxicity from frequent pesticide applications. However, the system requires careful management and, more importantly, efficient centralized production units which are expensive to build and operate. Thus, unless pesticide resistance becomes insurmountable, the release program will not be widely adopted commercially (Hussey and Bravenboer, 1971).

The Fillmore Citrus Protective District in Ventura County, California, has been using supplementary releases of natural enemies for biological pest control for almost half a century (DeBach, 1974). The district, which consists of 8,214 acres, maintains its own insectaries to annually produce predators and parasites for release as needed in orange and lemon groves. In 1973, supplementary releases included 9,674,000 of the encyrtid parasite, *Metaphycus helvolus* (Compere), for control of black scale, *Saissetia oleae* (Olivier); 28,193,000 of the aphelinid parasite, *Aphytis melinus* DeBach, for control of California red scale, *Aonidiella aurantii* (Maskell); 18,000 of the encyrtid parasite, *Leptomastix dactylopii* Howard; and 807,000 of the coccinellid predator, *Cryptolaemus montrouzieri* Mulsant, for control of mealybugs on citrus. Pesticide treatments are used only as necessary and are restricted where possible to highly selective materials. In 1973, less than 1.0 percent of the District's acreage was sprayed for California red scale and only about 8.0 percent for black scale.

The above examples of commercial use of supplementary releases exemplify what can be done towards manipulating pest and natural enemy populations for biological pest control. In both cases, careful management is a key factor in their success.

Literature Cited

DeBach, P.
1974. Biological Control by Natural Enemies. Cambridge Univ. Press, New York. 323 pp.

DeBach, P. and K. S. Hagen.
1964. Manipulation of entomophagous species. pp. 429-58. *In* P. DeBach (ed.) Biological Control of Insect Pests and Weeds. Reinhold Publ. Corp., New York. 844 pp.

Doutt, R. L. and K. S. Hagen.
1949. Periodic colonization of *Chrysopa californica* as a possible control of mealybugs. J. Econ. Entomol. 42(3): 560.

Huffaker, C. B. and C. E. Kennett.
1956. Experimental studies on predation: predation and cyclamen mite populations on strawberries in California. Hilgardia 26(4): 191-222.

Hussey, N. W. and L. Bravenboer.
1971. Control of pests in glasshouse culture by the introduction of natural enemies, pp. 195-215. *In* C. B. Huffaker (ed.) Biological Control. Plenum Press, New York. 511 pp.

Lingren, P. D.
1970. Biological control—can it be effectively used in cotton production today? Proc. 2nd Ann. Texas Conf. on Insect, Plant Disease, Weed and Bush Control. Texas A & M Univ., College Station. pp. 236-40.

McMurtry, J. A., H. G. Johnson, and G. T. Scriven.
1969. Experiments to determine effects of mass releases of *Stethorus picipes* on the level of infestation of the avocado brown mite. J. Econ. Entomol. 62(5): 1216-21.

Oatman, E. R., J. A. McMurtry, and V. Voth.
1968. Suppression of the two-spotted spider mite on strawberry with mass releases of *Phytoseiulus persimilis*. J. Econ. Entomol. 61(6): 1517-21.

Oatman, E. R. and G. R. Platner.
1971. Biological control of the tomato fruitworm, cabbage looper, and hornworms on processing tomatoes in southern California, using mass releases of *Trichogramma pretiosum*. J. Econ. Entomol. 64(2):501-6.

Parker, F. D., F. R. Lawson, and R. E. Pinnell.
1971. Suppression of *Pieris rapae* using a new control system: mass releases of both the pest and its parasites. J. Econ. Entomol. 64(3): 721-35.

Ridgway, R. L.
1972. Use of parasites, predators, and microbial agents in management of insect pests of crops, pp. 51-62. *In* Implementing Practical Pest Management Strategies, Proc. Natl. Exten. Insect Pest Management Workshop, Purdue Univ. 206 pp.

Ridgway, R. L. and S. L. Jones.
1969. Inundative releases of *Chrysopa carnea* for control of *Heliothis* on cotton. J. Econ. Entomol. 62(1): 177-80.

Ridgway, R. L., R. E. Kinzer, and R. K. Morrison.
1974. Production and supplemental releases of parasites and predators for control of insect and spider mite pests of crops, pp. 110-116. *In* F. G. Maxwell and F. A. Harris (eds.) Proceedings of the Summer Institute of Biological Control of Plant Insect and Diseases. University and College Press (Mississippi), Jackson. 647 pp.

Chapter 7.

USE OF SUPPLEMENTAL FOODS AND BEHAVIORAL CHEMICALS TO INCREASE THE EFFECTIVENESS OF NATURAL ENEMIES

K. S. Hagen and G. W. Bishop

Introduction

"Too little" or "too late" is an expression often heard in criticism of the natural enemies system for not controlling certain pest populations, and this is often valid with regards to parasites and predators of pests that attack monoculture crops. Natural enemies which become disassociated from their prey either in space or time may arrive and reproduce too late to prevent pests from attaining damaging levels. Disassociation of pests from their natural enemies can occur because of agricultural crop isolation from natural enemy reservoirs or refuges, and also from the use of non-selective insecticides which either temporarily remove natural enemies or reduce their prey or hosts. Some predators can also cause such depressions of the pest populations that not enough prey are left to retain them, and the predators are forced to leave.

As a rule, predators require feeding in the adult stage in order to produce eggs. The prey of the adult predator is not always the same species as the prey of the immature predator. Also, some are not predaceous as adults and must obtain other food such as nectar, pollen or honeydew before they can oviposit and reproduce. These adult foods may not be available in or around crops which harbor potential prey for predaceous larvae and nymphs. Some adult parasitic Hymenoptera and Diptera also rely on foods other than their hosts before becoming reproductively active. Monocultures and clean culture frequently eliminate necessary adult food even if prey for immature predators and parasites is present.

Any manipulations in the field that increase the chances of natural enemies finding their prey or hosts certainly increase the effectivess of natural enemies. There are some preliminary field experiments indicating that searching by natural enemies can be manipulated by use of kairomones. Kairomones are interspecific chemical signals that benefit the receiver organisms. In this chapter, we shall briefly deal with the use of supplementary foods for predators in the field as well as the use and role of kairomones that influence predators and parasites.

Supplementary Feeding

Predators or parasites can be manipulated and their effectiveness increased by providing artificial foods to either retain, arrest or stimulate them to oviposit (Hagen *et al.*, 1971: Hagen and Hale, 1974: Hagen, 1976; Ben Saad and Bishop 1976*a,b*). Predators relying strictly on living prey to reproduce can also be manipulated by conserving or adding target or alternate prey. Alternate hosts for parasites can also be manipulated. We do not yet have satisfactory artificial prey or hosts that can be placed in crops to simulate completely living prey or hosts, but undoubtedly this will become a reality in the future.

Artificial diets

The artificial diets that have been used to manipulate natural enemies vary from simple sugar solutions to nutritionally complete foods that may or may not include an attractant or a kairomone. The complexity of diet to be used depends upon the biology of the predators or parasites to be manipulated and whether they are to be retained, arrested, attracted, or stimulated to produce eggs and oviposit ("retaining" = keeping them in an area; "arresting" = stopping them on the way through). Some supplemental foods used in the field can induce all of these responses in certain predators (Hagen *et al.*, 1971; Hagen and Hale, 1974).

Retaining or arresting natural enemies. In crops where natural enemies are present but will probably disperse or starve because prey populations have decreased to low levels, it may be feasible to supply an artificial diet to retain or arrest them until the target prey or host pest populations begin to increase. By simply spraying sugar or molasses solutions on various crops, certain predators have been retained. Spraying a 10 percent sucrose solution on corn plants resulted in increases of coccinellid adults, chrysopid adults and eggs, and more effective suppression of corn leaf aphid populations in corn fields (Ewert and Chiang, 1966).

In a later experiment, where either sucrose or molasses solutions were sprayed on corn plants, chrysopid populations increased and this in turn reduced the number of European corn borer larvae before they entered the corn stalks (Carlson and Chiang, 1973). These sprays also increased the number of four-spotted fungus beetles, *Glischrochilus quadrisignatus* (Day), which entered the stalks and killed corn borer larvae. However, neither the chrysopids nor the fungus beetles responded to sucrose or molasses sprays when corn leaf aphid populations were high.

Coccinellids and chrysopids are not attracted to the sucrose but are arrested and aggregate as they contact the sucrose during random searching. This has been demonstrated by spraying sucrose on alfalfa and covering the sprayed areas with sticky screen trap cages. As many coccinellid or chrysopid adults were trapped on cages covering untreated alfalfa as were trapped on the cages covering sucrose (Hagen *et al.*, 1971). In alfalfa plots sprayed with a sucrose solution in June, when very few aphids were present, *Chrysopa carnea* Stephens and *Hippodamia* spp. adult numbers were greatly increased by arresting these insects as they contacted food in their random search. One day before spraying a sugar solution of 350 g sucrose plus 1000 ml water in alfalfa plots measuring 2 x 70 m, about 0.01 *C. carnea* adults and 0.2 coccinellid adults were taken per sweep. The day following the application of sucrose, about two *C. carnea* adults and four coccinellid adults were taken per sweep (Hagen *et al.*, 1976).

Attractants added to supplementary foods. Non-predaceous *C. carnea* adults have been manipulated in the field through the use of Wheast mixed with sugar to produce an artificial honeydew. Wheast is the yeast *Kluyreromyces fragilis* plus milk-whey substrate (Hagen and Tassan, 1970; Hagen *et al.*, 1971; Butler and Ritchie, 1971; Ben Saad and Bishop, 1972, 1976 *a,b) and acid hydrolysates of casein and beef blood* (Hagen *et al.*, 1976). Syrphid adults and their ichneumonid larval parasite *Diplazon* spp. are also attracted to artificial honeydews (Hagen *et al.*, 1971).

The main attractant for *C. carnea* in the above proteins is tryptophan. The mixture of 1 g tryptophan in 60 ml HCL plus 350 g sucrose plus 1000 ml H_2O was sprayed in alfalfa plots measuring 2 x 70 m. Where populations of *C. carnea* adults were about .02 adults per sweep (87 per acre) prior to treatment, more than four adults per sweep were taken in the tryptophan treatments in 24 hours as compared to two per sweep in plots sprayed with sucrose alone. This indicated that tryptophan attracted *C.carnea* adults from about 2.6 acres into 0.001 acre area treated (Hagen *et al.*, 1976).

Ben Saad and Bishop (1976*a*) found that potato plants sprayed with molasses, honey and tryptophan, either alone or combined with Wheast, attracted adults of *C. carnea, Hippodamia* spp., *Coccinella tranversoguttata richardsonis* Brown, *Scymnus postpinctus* Casey, and *Georcoris pallens* Stål and the pest *Lygus hesperus* Knight. Honey and tryptophan were the most attractive to all species. Hagen *et al.* (1976) did not find *L. hesperus* to be attracted to food sprays applied in alfalfa fields. However, migrating *Lygus* adults are sometimes arrested by sucrose (Linquist and Sorensen, 1970) and this has been also observed with some food sprays (Hagen and Hale, 1974).

In alfalfa fields there does not seem to be distinct attraction of coccinellids to the food sprays (Hagen *et al.*, 1971, 1976), but Ben Saad and Bishop (1976*a*) apparently obtained attraction of several *Hippodamia* spp. and *S. postpinctus* to honey, while *C. transversoguttata richardsoni* and *Hippodamia sinuata sinuata* Mulsant appeared to be attracted to aphid juice. Attraction was shown in the experiments by using trap cages covering the attractants tested.

Chemicals emanating from prey or hosts which attract predators or parasites are considered kairomones. Therefore, tryptophan or indole acetaldehyde could be called kairomones. As kairomones which influence natural enemies are discovered, they could be combined with supplementary foods applied in the field to increase attraction or to influence searching behavior.

Stimulating egg production and oviposition in the field. Most predaceous adult insects require a complex diet to produce eggs, while many phytophagous insects—especially among Lepidoptera—do not (Hagen, 1976). For example, among the aphidophagous predators, coccinellid adults require aphids to produce eggs, syrphid adults require pollen and of 30 species of *Chrysopa*, in which the adult feeding habits are known, 15 species require aphids to produce eggs. The other 15 species of *Chrysopa* apparently feed on honeydews, nectars, and pollen as adults (Hagen *et al.*, 1970). Adults of *C. carnea* carry over enough metabolites from their larval feeding to produce a total of about 30 eggs, but when fed complex artificial diets containing proteins which simulate honeydews they are able to deposit about 30 eggs per day, often over 1000 eggs in a lifetime (Hagen, 1950; Hagen and Tassan, 1970). Because *C. carnea* (fig. 13) is one of the few predators for which relatively inexpensive artificial diets have been developed to stimulate egg production, it is about the only species to date that has been manipulated in the field to increase its oviposition. *C. carnea* apparently does not require aphids to oviposit. Even though they are attracted to artificial honeydews many common syrphid species need aphid colonies to stimulate oviposition (Ben Saad and Bishop 1976a, Hagen *et al.*, 1971). *Hippodamia* spp. oviposition was increased in alfalfa fields by spraying artificial honeydews if there were a few aphids present to supplement the artificial diet (Hagen *et al.*, 1971).

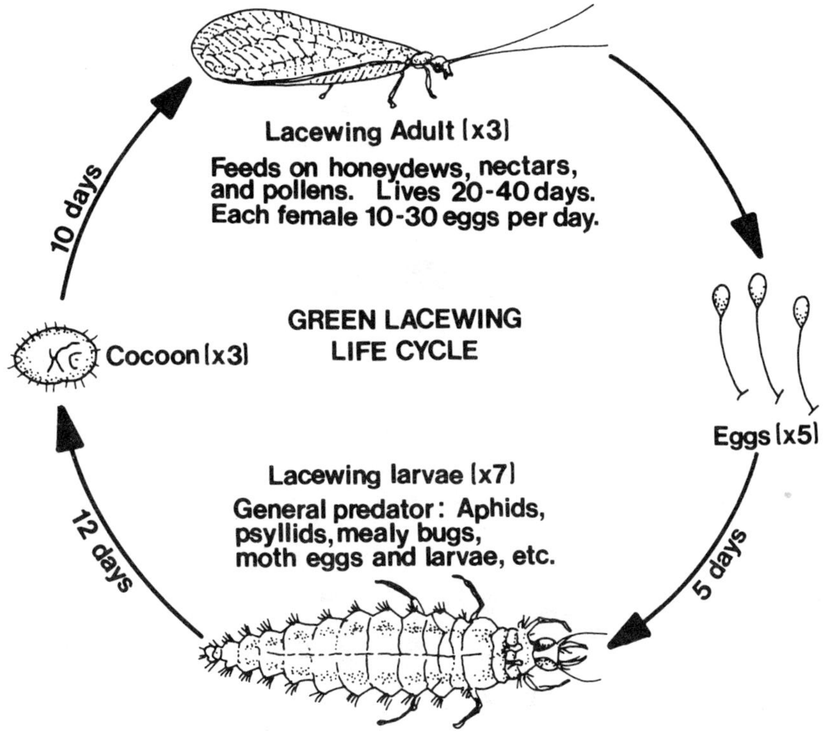

Fig. 13. Life cycle of *Chrysopa carnea* at 24°C or 75°F. (Tassan and Hagen, 1970).

By spraying fields with artificial honeydews high aphid populations are simulated and *C. carnea* is attracted to the food sprays and oviposits (figs. 14, 15). It is possible to obtain high *C. carnea* larval populations in crops much earlier than under natural conditions.

Spraying yeast hydrolysate plus sugar plus water (4:7:10 parts respectively), or Wheast plus sugar plus water (3:3:30 parts respectively), reduced aphid populations to less than one-half the population density as compared to controls. This was shown in alfalfa (Hagen *et al.* 1971; fig. 16) and in green peppers (Hagen and Hale, 1974; fig. 17). *Myzus persicae, Lygus hesperus* and *Autographa californica* (Speyer) numbers were reduced in potato fields by using a Wheast plus sugar solution (Ben Saad and Bishop, 1976*a*).

Because *C. carnea* adults can be attracted to fields where artificial honeydew is applied, the target prey of *Chrysopas* larvae may be insects other than aphids because the larvae are general predators. Spraying cotton plots in California with Wheast plus sugar increased *Chrysopa* egg deposition at least three-fold, and bollworm eggs, larvae and damaged bolls were reduced in treated plots (fig. 18; Hagen *et al.*, 1970, 1971). In Arkansas, borders of a cotton field were sprayed with a Wheast and sugar solution three times in early spring; later the field was sprayed in alternate strips 45 feet wide by aircraft using 10 gallons (1 pound Wheast plus 1 pound sugar in 1 gallon water) per acre. The cotton yield in the Wheast-treated field was 1204 pounds per acre, and that from an adjacent pesticide-treated field was 1136 pounds per acre (Denver, 1974).

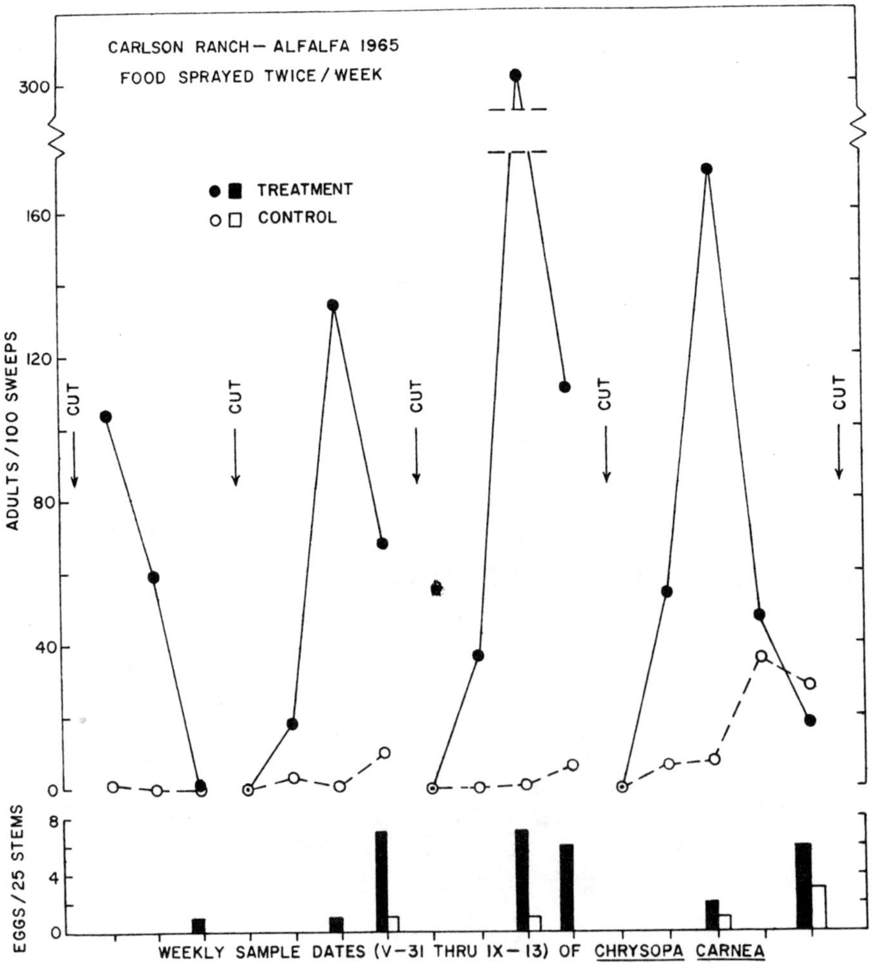

Fig. 14. Mean number of *Chrysopa carnea* adults and eggs sampled during the growing season in an alfalfa plot sprayed twice weekly with a spray of yeast hydrolysate plus sugar plus water (4:7:10) compared to the number of adults and eggs of *C. carnea* sampled in a control plot in the same field. (Hagen and Tassan, 1971).

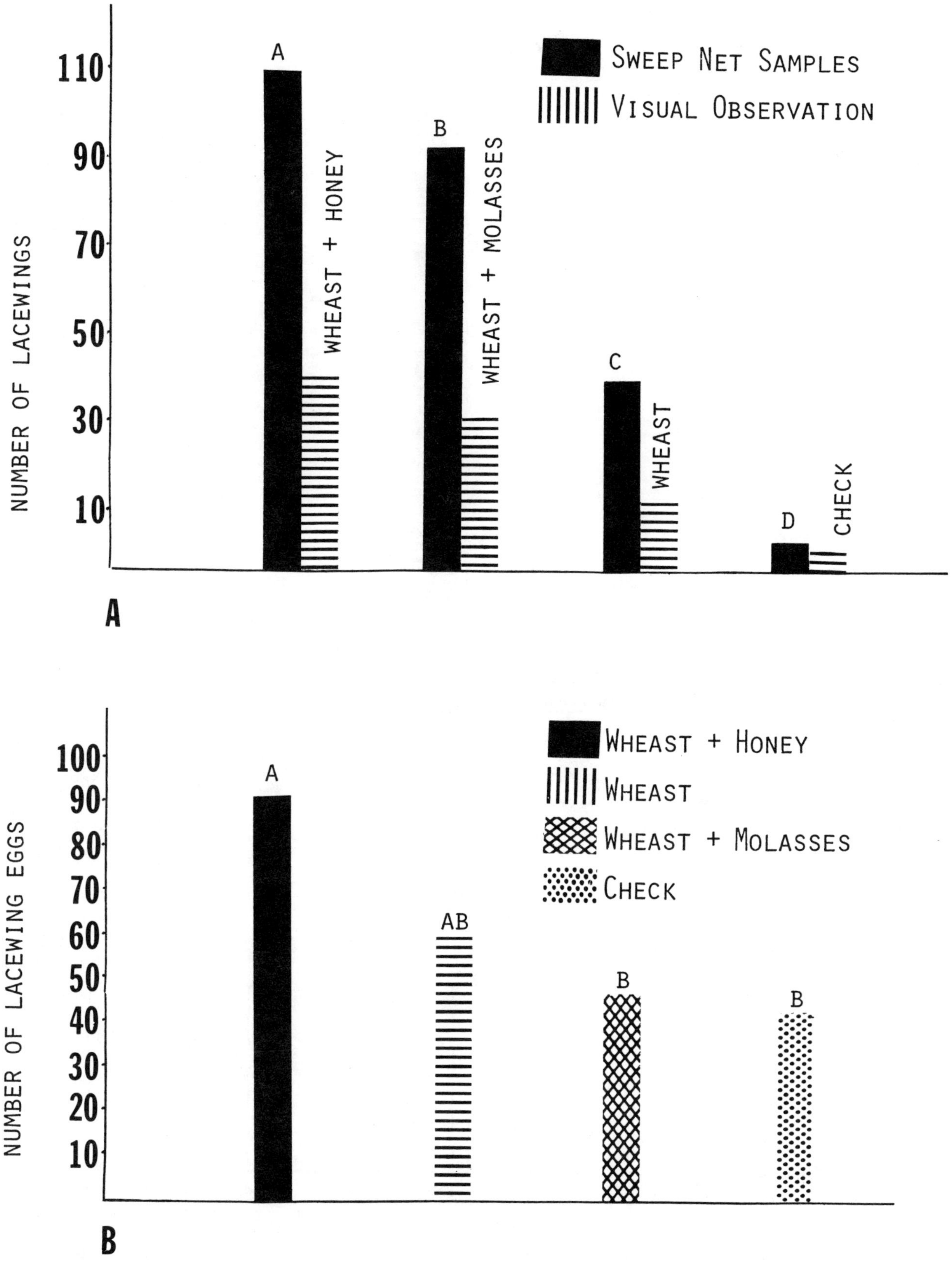

Fig. 15. Total number of (A) *C. carnea* in 1100 sweeps and in a 15-minute visual count in 45-m² area and (B) *C. carnea* eggs in 1000 potato leaves. Parma, Idaho, 1972. Bars with the same letter are not significantly different by DMRT at 5 percent level. (Ben Saad and Bishop, 1976*a*).

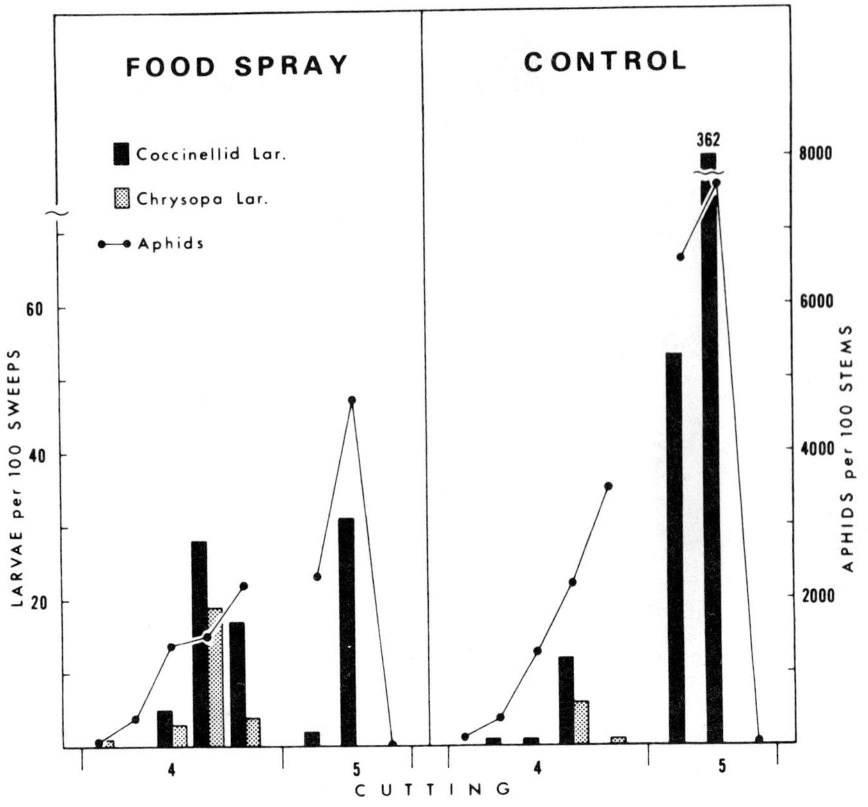

Fig. 16. Mean number of coccinellid larvae (mainly *Hippodamia* spp.) and *Chrysopa carnea* larvae and aphids (mainly spotted alfalfa aphids) sampled in an alfalfa plot sprayed once a week in 1968 with yeast hydrolysate plus sugar plus water (4:7:10) compared to the number of coccinellid and *Chrysopa* larvae and aphids sampled in a control plot in the same field. (Hagen *et al.*, 1971).

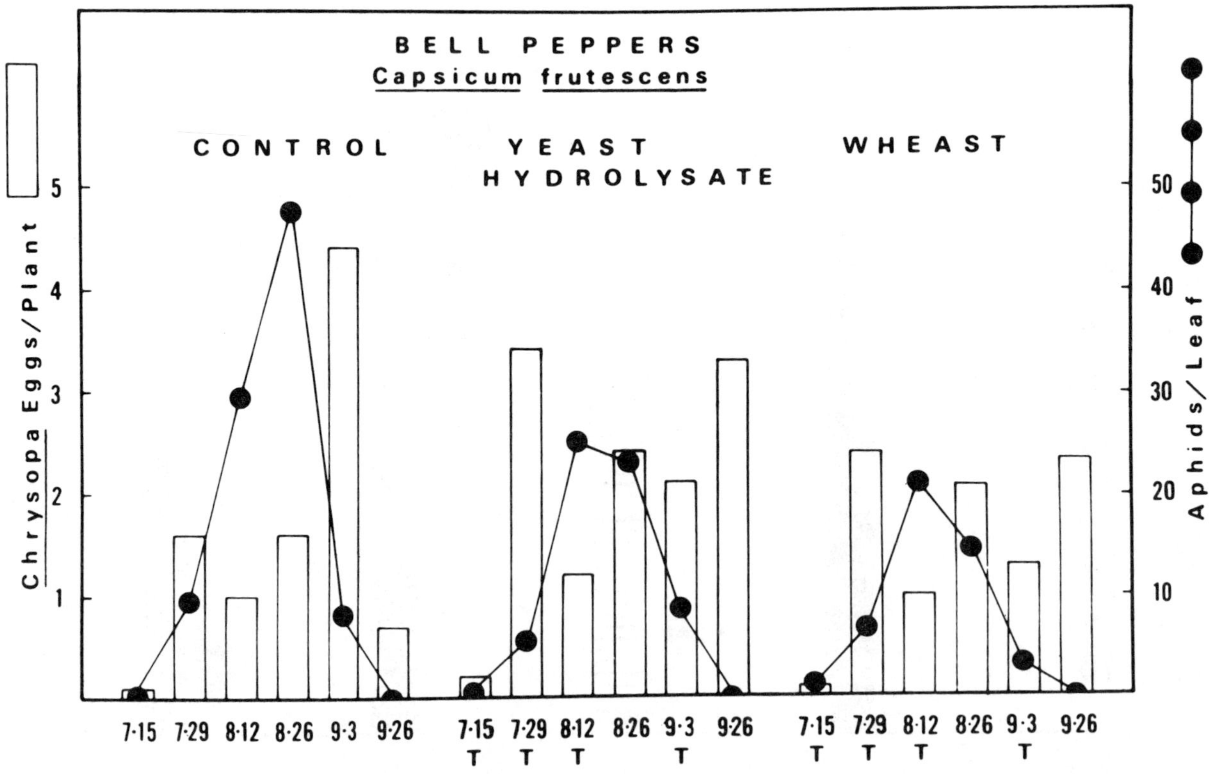

Fig. 17. Influence of two types of predator food sprays applied to bell peppers on *Chrysopa carnea* egg deposition and abundance of green peach aphids. (The T's beneath the dates show when food sprays were made). (Hagen and Hale, 1974).

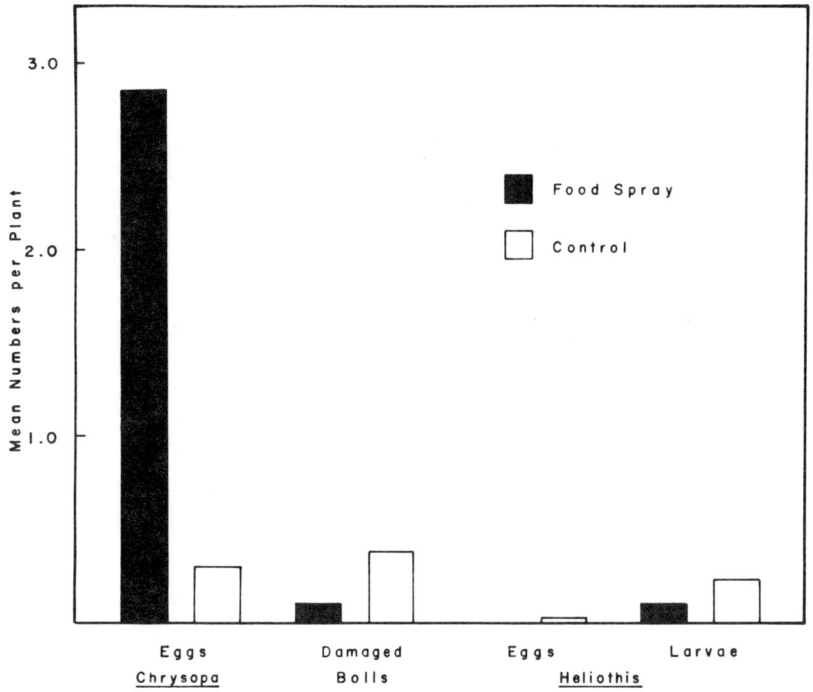

Fig. 18. Mean number of *Chrysopa carnea* eggs, bollworm eggs, larvae and damaged bolls in a plot sprayed with Wheast plus sugar plus water (450g plus 450g plus 3500ml) in September and sampled 1 week later compared to those sampled from control plot in the same field. (Hagen *et al.,* 1970).

Even where sucrose sprays arrest *C. carnea* some eggs may be deposited because of prior feeding in other areas, but to increase egg production it is necessary to provide diets containing protein. This was clearly shown when tryptophan plus sucrose sprayed in an alfalfa field attracted as many *Chrysopa* as did complete diets containing Wheast. Tryptophan plus sugar alone was not a complete enough diet for egg production.

Wheast products with 10 percent ash or salt are not being produced commercially. Dried yeasts plus whey products (casein protein) plus sugar and water diets (not more than 10 percent salts) have stimulated egg production of *C. carnea* in the laboratory to about the same level as did Wheast plus sugar and water (Hagen, unpublished data), but these artificial honeydew diets are just being field-tested. Recent field tests indicate that formula 57 can be a substitute for Wheast, which is no longer commercially available.

Natural diets

Natural products such as pollen, nectar, and plant seeds have been used in experimental plots to increase the effectiveness of some predatory insects and mites. The utilization of wild flowers and weeds as a source of pollen and nectar by parasites has also been demonstrated.

Wild pollen and nectar. The presence of cover crops, adjacent crops, ornametals and weeds is often beneficial to natural enemies of insects and spider mites (van Emden 1965; Southwood and Way, 1970; DeLoach, 1971). These plants may harbor alternate prey and also provide pollen and nectar food for a variety of parasites and predators. A few studies have been made on the influence of wild vegetation upon insects in relation to crops.

Fifteen unsprayed apple orchards in Canada were classified on the basis of the relative abundance of nectar-producing flowers in the undergrowth as: rich, average or poor. Proportionately, about 18 times as many tent caterpillar pupae were parasitized in orchards having rich undergrowths of wild flowers as were parasitized in orchards having poor floral undergrowths. Parasitism of tent caterpillar eggs was increased four times and that of codling moth larvae five times. Orchards having an average undergrowth of wild flowers produced intermediate values (Leius, 1967).

The control of certain weeds has reduced the impact of some tachinid and hymenopterous parasites. Shortly after the accidental introduction of the New Guinea sugarcane weevil, a tachinid parasite, *Lixophaga spenophori* (Villeneuve), was imported into Hawaii. The weevil came under biological control by this parasite, but in the last years the extent of control decreased and the decrease is believed to be correlated with changes in cultural practices in sugarcane fields (Beardsley, 1968). The tachinid adults were observed to feed on nectar from floral and extra-floral nectaries of certain weed species of Euphorbiaceae. The reduction of these weeds and possible toxicity of herbicides to the tachinids (fig. 19) could explain the increase of weevil damage (Leeper, 1974). Topham and Beardsley's (1975) studies on this problem corroborate Leeper's findings. However, these authors are not recommending complete abandonment of field-edge weed control because faster growing weeds crowd out desirable *Euphorbia* species. Topham and Beardsley suggest that. for maximum utilization of the biological control potential of the tachinid, there is need for development of a herbicide program utilizing materials and application intervals which will favor the continuous maintence of field-margin stands of nectar-producing *Euphorbia* species.

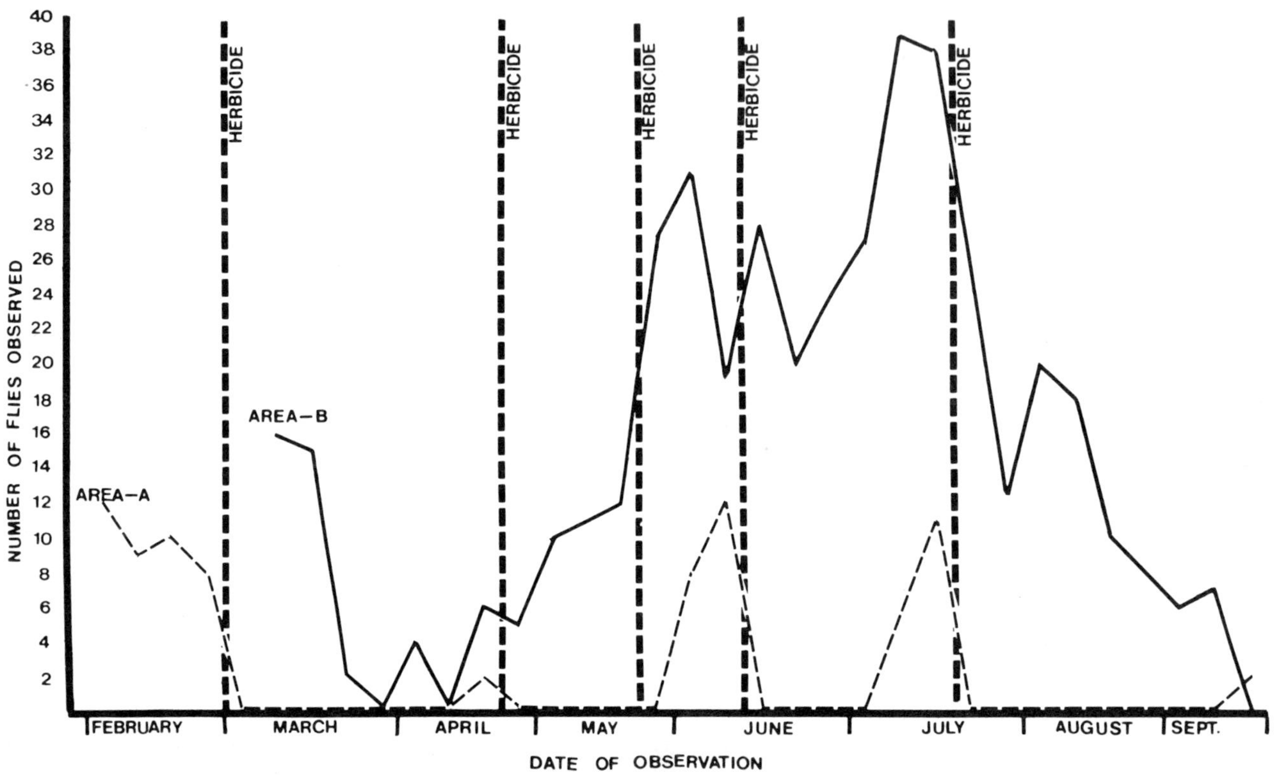

Fig. 19. Comparison of numbers of adult *Lixophaga sphenophori* observed on weeds in Area A (herbicide treated) and Area B (untreated) during 1972. (From Topham and Beardsley, Jr. 1975).

In Hawaiian sugar cane fields, two hymenopterous parasites of scarab grubs were observed feeding on nectar of several weeds. The scoliid *Campsomerus marginella modesta* Smith and the tiphiid *Tiphia segregata* Crawford were observed to feed on flowers of six weed species belonging to six plant families. *Euphorbia geniculata* flowers visited by the *Lixophaga* also served as a nectar source of both scarab parasites (Beardsley personal communication).

Various pollens are important alternate foods for certain predatory mites. Some predaceous phytoseiid mites like *Ambylseius hibisci* (Chant) can be cultured continuously on pollens alone, but *Typhlodromus pyri* Scheuten can develop only one generation and *Typhlodromus occidentalis* Nesbitt does not feed at all on pollens (Huffaker *et al.*, 1970). *Ambylseius fallacis* (Garman) fed pollen from the trumpet creeper vine and the tulip poplar grew and survived as well as when fed twospotted spider mite eggs (Ahlstrom and Rock, 1973).

Pollens vary in their chemical composition and not all pollens are effective foods. C. E. Kennett (personal communication) exposed *A. hibisci* to different pollens from different native plants including both anemophilous (wind-borne) and entomophilous (insect-borne) type pollens. About one-half of the species tested in each of those types allowed two generations of the phytoseiid mite to develop. Among the most desirable pollens for *A. hibisci* were cattail (*Typha*), oak, and willow pollens; pollens from pines or olive were not effective.

Pollen dusting. Pollens have been dusted on plants to increase the effectiveness of some phytoseiid mites, and also to increase populations of nontarget scavenger mites which serve as an alternate prey source when target pest mite populations are low.

McMurtry and Scriven (1966) found that *A. hibisci* controlled the avocado brown mite when pollen was added, but not where pollen was omitted.

Tydeid mites feed on honeydews and pollen in nature, and can serve as an alternate prey for the strictly predaceous *T. occidentalis* when the Willamette mite is scarce in vineyards. To increase the amount of tydeid mites, Flaherty and Hoy (1971) and Calvert and Huffaker (1974) treated vines with *Typha* pollen and increased the tydeid mites which in turn significantly increased the abundance of *T. occidentalis* and the Willamette mite remained low the rest of the season (Flaherty *et al.*, 1971).

Seeds. There are many general predators, particularly in the Heteroptera, that feed on plants to obtain mainly water. Some of these feed on seeds without causing conspicuous plant damage. *Geocoris bullatus* (Say) and *G. pallens* developed faster and produced more eggs when reared on a combination diet of green plants, insect prey and sunflower seeds than they did when feeding on any of these dietary components alone (Tamaki and Weeks, 1972). These authors scattered ¼ pounds of chopped sunflower seeds on 180 square feet twice in replicated sugar-beet field plots to determine if the supplementary feeding would increase the *Geocoris* population. The treated plots had over twice as many *Geocoris* eggs as did plots without sunflower seeds. *Geocoris* are general predators, so any increase of these bugs would help prevent outbreaks of aphids. lepidopterous eggs and spider mites.

Behavior-Modifying Chemicals

In recent years much research has been devoted to isolating, identifying and applying behavioral chemicals that influence insects, and in 1970 books dealing with behavioral chemicals were published. The application of behavioral chemicals, mainly pheromones, in insect control has been reviewed by Birch (1974) and Chambers (1973), and of particular interest to the present paper is a review on the responses of parasitic insects to behavioral chemicals (kairomones) in locating hosts (Lewis *et al.*, 1976) .

The classification of behavioral chemicals is evolving and not all researchers agree with some of the terms used (Nordlund and Lewis, 1976). The terms "hormone" and "pheromone" have posed no problem, but the terms "allomone" and "kairomone," interspecific communication chemicals, are not yet universally accepted. Allomones are interspecific, behavioral chemicals that favor the sender and kairomones are those that benefit the receiver.

Sources of chemical cues used by natural enemies to find their hosts

Predatory and parasitic insects have evolved the ability of responding to certain chemicals emanating from their environment that lead them to finding and utilizing their prey or hosts. There is a wide variety of organic chemicals involved and these chemicals can range from distant attractants, generally perceived in flight by natural enemies, to chemicals contacted directly on or in the insect host itself .

The chemical attractants, kairomones, that attract natural enemies from a distance of at least several meters are used by some natural enemies to find the habitat which may harbor their prey or hosts. Certain tachinid flies and ichneumonids have been shown to be attracted to plants that support their hosts (Monteith 1958: Arthur, 1962: Reed *et al*, 1970, and others). The large coccinellid, *Anatis ocellata* (Linn.), is attracted to the volatile chemicals associated with conifer trees rather than to their aphid prey (Kesten, 1969).

Pheromones secreted by potential host insects attract certain natural enemies from a distance. Predaceous flies and beetles and certain parasites of bark beetles employ this tactic, and some of them respond to terpenes of attacked trees (Rice, 1969: Wood, 1970: Pitman and Vite, 1971; Williamson, 1971; Camors and Payne, 1972; Dyer, 1975). Certain tachinid parasites of pentatomid bugs orient to male-bug pheromones and oviposit mainly on males (Mitchell and Mau, 1971; Todd and Lewis, 1976).

Frass, feces and damaged plant tissue at feeding sites of potential hosts or prey emit chemical stimuli which are perceived by many natural enemies. Some of these chemicals have been identified. Some have been found in mandibular glands, others are associated with symbiotic organisms of the phytophagous hosts. These kairomones are often perceived in flight by natural enemies (Lewis *et al.*, 1976). Honeydew excreted by aphids and mealy-bugs apparently contains a breakdown product of tryptophan which is a powerful attractant of *Chrysopa carnea* (Hagen *et al.*, 1976).

There are chemicals associated with the cuticle and hemolymph of host insects that influence the behavior of natural enemies in close proximity to their prey or hosts. Lewis *et al.* (1972) observed that scales left by ovipositing female *Heliothis zea* (Boddie) were a source of a kairomone that caused the random searching habit of *Trichogramma evanescens* Westwood to change to tight circles. The most active *Trichogramma* host-seeking kairomone was identified as tricosane, isolated from hexane extracts of moth scales (Jones *et al.*, 1973).

Several other kairomones that influence searching behavior of parasitic Hymenoptera have been chemically identified, but thus far apparently only tricosane has been tested in the field (Lewis *et al.*, 1976).

Application of kairomones in the field

Application of kairomones in the field to increase predation and parasitization by natural enemies has only recently begun. Use of tryptophan in sugar solutions or mixed with sugar plus protein foods to attract searching *Chrysopa carnea* adults into fields and to increase egg deposition has already been discussed. Apparently, the only parasites to be tested in the field with the use of kairomones are *Trichogramma* spp. Lewis *et al.* (1975) demonstrated that a hexane extract of *Heliothis zea* scales or tricosane applied in the field increased Trichogramma parasitization of *H. zea* eggs. The presence of the added kairomone continuously reinforced the intensified searching behavior for at least 3 days. Constant exposure to kairomones tends to increase the longevity and total production of progeny of *Trichogramma pretiosum* Riley, which resulted in increased parasitization of host eggs (Nordlund *et al.*, 1976). However, there should be a balance between host-egg density and coverage or frequency of kairomone treatments. Total coverage of a crop with a hexane moth scale extract, or tricosane in hexane, stimulates naturally occurring and/or released *Trichogramma* into an intensive searching behavior. Therefore, the total area searched by an individual parasite is reduced. In a high host-egg population, many host eggs could be parasitized but the number of host eggs found depends upon how many *Trichogramma* are present. Lewis and his colleagues are now applying various densities of kairomone-treated inert particles in the field to turn on and off the intensified searching behavior for maximum coverage of the crop with the least number of *Trichogramma* released. Irradiated host eggs rendered nonviable still allow *Trichogramma* to develop. These eggs together with kairomones have been thought to increase *Trichogramma* populations early in fields.

Conclusion

Increasing predation and parasitization through use of supplementary foods and kairomones has been demonstrated in the field. Although application of these techniques is still mainly in an experimental phase, there is much promise in their use as additional tools to be employed by pest managers where naturally-occurring biological control agents require augmentation. They can be used in conjunction with releases of natural enemies.

Literature Cited

AHLSTROM, K. R. and G. C. ROCK.
1973. Comparative studies on *Neoseiulus fallacis* and *Metaseiulus occidentalis* for azinphosmethyl toxicity and effects of prey and pollen on growth. Ann. Entomol. Soc. Amer. 66: 1109-13.

ARTHUR, A. P.
1962. Influence of host tree on abundance of *Itoplectis conquisitor* (Say), a polyphagous parasite of the European pine shoot moth, *Rhyacionia buoliana* (Schiff.). Can. Entomol. 94: 337-47.

BEARDSLEY, J. W., JR.
1968. A proposal for research on sugarcane insects; evaluation and augmentation of natural enemies of sugarcane pests in Hawaii. Paper submitted to H.S.P.A. (unpublished).

BEN SAAD, A. A. and G. W. BISHOP
1972. The use of food sprays to augment the natural enemies of potato insects. Abstracts 14th Inter. Congr. Entomol.
1976a. Effect of artificial honeydews on insect communities in potato fields. Environ. Entomol. 5: 453-57.
1976b. Effect of artificial honeydews on numbers of insects in potato fields. Univ. of Idaho, Dept. of Entomol. Anniversary Publ. (4: 23-25.)
1976c. Attraction of insects to potato plants through use of artificial honeydews and aphid juice. Entomophaga 2: 49-57.

BIRCH, M. C., ed.
1974. Pheromones. North Holland Publ. Co., Amsterdam, London, 495 pp

BUTLER, G. D., JR. and P. L. RITCHIE, JR.
1971. Feed wheast and the abundance and fecundity of *Chrysopa carnea.* J. Econ. Entomol. 64: 933-34.

CALVERT, D. J. and C. B. HUFFAKER.
1974. Predator *(Metaseiulus occidentalis)* under sulfur and cattail pollen applications in a non-commercial vineyard. Entomophaga 19: 361-69.

CAMORS, F. B., JR. and T. L. PAYNE.
1972. Response of *Heydonia unica* (Hymenoptera: Pteromalidae) to *Dendroctonus frontalis* (Coleoptera: Scolytidae) pheromones and a host-tree terpene. Ann. Entomol. Soc. Amer. 65: 31-33.

CARLSON, R. E. and H. C. CHIANG.
1973. Reduction of an *Ostrina nubilalis* population by predatory insects attracted by sucrose sprays. Entomophaga 18: 205-11.

CHAMBERS, D. L.
1973. Chemical messengers as tools in the management of insect pests. Proc. Tall Timbers Conf. Ecol. Animal Contr. Habitat Management (1973) No. 5: 135-45.

DELOACH, C. J.
1971. The effect of habitat diversity of predation. Proc. Tall Timbers Conf. Ecol. Animal Contr. Habitat Management (1970) No. 2: 223-39.

DENVER, C.
1974. The use of biological control in an Arkansas farming operation. Proc. Tall Timbers Conf. Ecol. Animal Contr. Habitat Management (1973) No. 5: 61-65.

DYER, E. D. A.
1975. Frontalin attractant in stands infested by the spruce beetle, *Dendroctonus rufipennis* (Coleoptera: Scolytidae). Can. Entomol. 107: 979-88.

EWERT, M. A. and H. C. CHIANG.
1966. Dispersal of three species of coccinellids in corn fields. Can. Entomol. 98: 999-1003.

FLAHERTY, D. L. and M. A. HOY.
1971. Biological control of Pacific mites in San Joaquin Valley vineyards: Part III. Role of tydeid mites. Res. Popul. Ecol. 13: 80-96.

FLAHERTY, D. L., C. LYNN, F. JENSEN and M. HOY
1971. Influence of environment and cultural practices on spider mite abundance in southern San Joaquin Valley Thompson seedless vineyards. Calif. Agr. 25(11): 6-8.

HAGEN, K. S.
1950. Fecundity of *Chrysopa californica* as affected by synthetic foods. J. Econ. Entomol. 43: 101-04.

1976. Role of nutrition in insect management. Proc. Tall Timbers Conf. Ecol. Animal Cont. Habitat Management 6: 221-61.

HAGEN, K. S., P. GREANY, E. F. SAWALL, JR. and R. L. TASSAN.
1976. Tryptophan in artificial honeydews as a source of an attractant for adult *Chrysopa carnea.* Environ. Entomol. 5: 458-68.

HAGEN, K. S. and R. HALE.
1974. Increasing natural enemies through use of supplementary feeding and non-target prey. pp. 170-181 in Proceedings of the Summer Institute Biological Control of Plant Insects and Diseases. F.C. Maxwell and F.A. Harris, eds. Univ. Press of Miss., Jackson.

HAGEN, K. S., E. F. SAWALL, JR., and R. L. TASSAN.
1971. The use of food sprays to increase effectiveness of entomophagous insects. Proc. Tall Timbers Conf. Ecol. Animal Contr. by Habitat Management (1970) No. 2: 59-81.
HAGEN, K. S. and R. L. TASSAN.
1970. The influence of Food Wheast® and related *Saccharomyces fragilis* yeast product on the fecundity of *Chrysopa carnea*. Can. Entomol. 102: 806-11.
HAGEN, K. W., R. L. TASSAN and E. F. SAWALL, JR.
1970. Some ecophysiological relationships between certain *Chrysopa*, honeydews and yeasts. Boll. Lab. Entomol. Agrar. Portici 28: 113-34.
HUFFAKER, C. B., M. VAN DE VRIE and J. A. MCMURTRY.
1970. Tetranychid populations and their possible control by predators. An evaluation. Hilgardia 49: 391-458.
JONES, R. L., W. J. LEWIS, M. BEROZA, B. A. BIERL and A. N. SPARKS.
1973. Host-seeking stimulants Kairomones) for egg parasite, *Trichogramma evanescens*. Environ. Entomol. 2: 593-96.
KESTEN, U.
1969. Morphologie und Biologie van *Anatis ocellata* (L.) Z. angew. Entomol. 63: 412-45.
LEEPER, J. R.
1974. Adult feeding behavior of *Lixophaga spenophori*, a tachinid parasite of the New Guinea sugarcane weevil. Proc. Hawaii. Entomol. Soc. 21: 403-12.
LEIUS, K.
1967. Influence of wildflowers on parasitization of tent caterpillar and codling moth. Can. Entolmol. 99: 444-46.
LEWIS, W. J., R. L. JONES, H. R. GROSS, JR. and D. A. NORDLUND.
1976. The role of kairomones and other behavioral chemicals in host finding by parasitic insects. Behavioral Biol. 16: 267-89.
LEWIS, W. J., R. L. JONES, D. A. NORDLUND and A. N. SPARKS.
1975. Kairomones and their use for management of entomophagous insects. I. Evaluation for increasing rates of parasitization by *Trichogramma* spp. in the field. J. Chem. Ecol. 1: 343-47
LEWIS, W. J., R. L. JONES and A. N. SPARKS.
1972. A host-seeking stimulant for the egg parasite *Trichogramma evanescens;* its source and a demonstration of its laboratory and field activity. Ann. Entomol. Soc. Amer. 65: 1087-89.
LINDQUIST, R. K. and E. L. SORENSEN.
1970. Interrelationships among aphids, tarnished plant bugs and alfalfas. J. Econ. Entomol. 63: 192-95.
MITCHELL, W. C. and R. F. L. MAU.
1971. Response of the female southern green stink bug and its parasite. *Trichopoda pennipes*, to male stink bug pheromones. J. Econ. Entomol. 64: 856-59.
MCMURTRY, J. A. and G. T. SCRIVEN.
1966. The influence of pollen and prey density on the number of prey consumed by *Amblyseius hibisci*. Ann. Entomol. Soc. Amer. 59: 147-49.
MONTEITH, L. G.
1958. Influence of host and its food plant on host finding by *Drino bohemica* Mesn. (Diptera: Tachinidae) and interaction of other factors. Proc. 10th Intern. Congr. Entomol. (1956) 2: 603-06.
NORLUND, D. A. and W. J. LEWIS.
1976. Terminology of chemical releasing stimuli in intraspecific and interspecific interactions. J. Chem. Ecol. 2: 211-20.
NORLUND, D. A., W. J. LEWIS, R. L. JONES, H. R. GROSS and K. S. HAGEN.
1976. Kairomones and their use for management of entomophagous insects: VI. A critical examination of the kairomones for the predator *Chrysopa carnea* Stephens at oviposition sites of *Heliothis zea* (Boddie). J. Chem. Ecol. (In press.)
PITMAN, G. B. and J. B. VITE.
1971. Predator-prey response to western pine beetle attraction. J. Econ. Entomol. 64: 402-04.
REED, D. P., P. O. FEENY and R. B. ROOT.
1970. Habitat selection by the aphid parasite *Diaeretiella rapae* (Hymenoptera: Braconidae) and hyperparasite *Chairips brassicae* (Hymenoptera: Cynipidae). Can. Entomol. 102: 1567-78.
RICE, R. E.
1969. Response of some predators and parasites of *Ips confusus* (Lee) to olfactory attractants. Contrib. Boyce Thompson Inst. 24: 189-94.
SOUTHWOOD, T. R. E. and M. J. WAY.
1970. Ecological background to pest management. Pp. 6-29. *In* R. L. Rabb and F. E. Guthrie, eds., Concepts of Pest Management. Proc. Conf. North Carolina State Univ., Raleigh, pp. 242
TAMAKI, G. and R. E. WEEKS.
1972. Biology and ecology of two predators, *Geocoris pallens* Stal and *G. bullatus* (Say). USDA Tech. Bul. 1446, 46 pp.
TASSAN, R. L. and K. S. HAGEN.
1970. Culturing green lacewings in the home and school. Univ. Calif. Cooperative Ext. Serv., Leaflet 2500. 2 pp (out of print)
TODD, J. W. and W. J. LEWIS.
1976. Incidence and oviposition patterns of *Trichopoda pennipes* (F.), a parasite of the southern green stink bug, *Nezara viridula* (L.). J. Georgia Entomol. Soc. 11: 50-54.
TOPHAM, M. and J. W. BEARDSLEY, JR.
1975. Influence of nectar source plants on the New Guinea sugarcane weevil parasite, *Lixophaga sphenophori* (Villeneuve) Proc. Hawaii. Entomol. Soc. 22: 145-54.
VAN EMDEN, H. F.
1965. The role of uncultivated land in the biology of crop pests and beneficial insects. Scientif. Hort. 17: 121-36.
VAN EMDEN, H. F. and K. S. HAGEN.
1976. Olfactory reactions of the green lacewing, *Chrysopa carnea* to tryptophan and certain breakdown products. Environ Entomol. 5: 469-73.
WILLIAMSON, D. L.
1971. Olfactory discernment of prey by *Medetra bistriata* (Diptera; Dolichopodidae). Ann. Entomol. Soc. Amer. 64: 586- 89.
WOOD, D. L.
1970. Pheromones of bark beetles. Pp. 301-16. In: Control of Insect Behavior by Natural Products. D. L. Wood, R. M Silverstein and M. Nakajima, eds. Academic Press, N.Y. and London.

Chapter 8.
CULTURAL PRACTICES IN PEST MANAGEMENT

G. W. Bishop, D. W. Davis and T. F. Watson

Introduction

Most cultural control practices evolved with agriculture, and until the advent of modern pesticides such practices were often the only practical means of suppressing pests on crops. The past popularity of these methods can be attributed to their economy and relative ease of implementation, as they were essentially a manipulation or variation of farm practices.

After having received relatively little attention since the discovery of organic pesticides there is now a renewed interest in cultural practices for pest control. These methods are largely preventive, and achieve control by disrupting the life cycle of the pest or by providing a more favorable habitat for parasites and predators. As cultural controls are effective only at limited phases of pest activity, timings can be critical. Development of cultural controls is best achieved after an intimate knowledge of the life cycle and development of the pest as related to all components of the agroecosystem. Conversely, cultural practices should not disrupt pest control.

Cultural controls are largely indirect, which makes their value difficult to evaluate. Levels of control are frequently below those of other control methods, and therefore many earlier practices were discontinued when synthetic organic pesticides appeared. These chemicals gave spectacular direct control of pests, but when their shortcomings as sole control agents were recognized, and concerns regarding their use increased, cultural methods took on a new importance. These methods are founded on ecological principles and are, therefore, compatible with modern philosophies of pest management. They lower general equilibrium levels of pest populations with minimum disruption of natural control factors.

Information on cultural practices is available from various sources, and some references are listed at the end of this paper. The following discussion will, therefore, emphasize principles and applications of the more recent studies, particularly those in project W-84.

Crop Management

Cultural practices normally available to the individual grower are those which can be directed against pests having limited powers of dispersal. Insects which spread readily must often be approached on an area-wide basis. Insects with long life cycles can be disrupted by techniques which destroy the available food or refuge sites at some critical point during development. Those insects with limited periods of dispersal or egg deposition can often be controlled by adjusting either planting or harvesting dates.

Crop rotation

Crop rotation consists of a planting scheme alternating susceptible and nonsusceptible crops. The necessary interval between susceptible crops depends upon the length of life cycle, reproductive potential, degree of specificity, and dispersal characteristics of the target pest. Pests most subject to this type of control are poorly mobile soil-inhabiting species with a restricted host range and a life cycle of 1 year or longer. This method provides control of species like the corn rootworms, which require specific hosts for oviposition and feeding, or wireworms and white grubs which require several seasons to develop.

Crop rotation continues to be an important method of pest control but is usually general in its effects, rather than being directed toward specific problems.

Planting and harvesting dates

Planting and harvesting dates of some crops may be altered to reduce or avoid damage from pests. Historically, delayed planting of wheat was the principal means of controlling the Hessian fly, *Mayetiola destructor* (Say), before resistant varieties were developed. Damage by the boll weevil, *Anthonomus grandis* Boheman, is commonly reduced by early planting to get bolls past the highly susceptible stage before weevils become abundant; early harvesting then reduces the number of overwintering weevils. Harvesting the first crop of alfalfa before damage by the alfalfa weevil, *Hypera postica* (Gyllenhal), becomes severe prevents or delays the need for insecticide treatment in Idaho and Utah. Harvesting dates for pest control in cotton (discussed later in this paper) and alfalfa received major attention in project W-84.

Trap crops

Trap crops attract and/or retain pest species, thereby reducing infestations in crops in the vicinity. Trap crops must either be more attractive than the crop to be protected or located in a position to interrupt movement of the pest into the crop. A system of alternating blocks of cotton with strips of alfalfa to attract and retain lygus bugs prevented damage in cotton (Rakickas and Watson, 1974). Research in progress indicates that, under some circumstances, treatment of trap crops with selective or systemic insecticides may improve effectiveness.

Strip harvesting

Natural enemies which have built up over a period of time are decimated when extensive acreages of alfalfa are harvested within a short time. This practice is especially disruptive in areas where other suitable habitats are meager. Research showed that strip cutting of alfalfa could be used to preserve parasites and predators of the spotted alfalfa aphid by herding them into uncut strips of alfalfa. When the remainder of the alfalfa was cut, the beneficial insects would go back to the trips cut earlier. If any insecticides were needed they could be applied to the stubble with little adverse effect on biological control (Schlinger and Dietrick, 1960). Rakickas and Watson (1974) showed that displaced natural enemies used the uncut strips for both shelter and prey—important predators could be maintained throughout the summer by this system. Predator populations declined rapidly by early summer when a solid cutting method was used.

Lygus bugs are among the most important pests on a number of crops. Most damage in southwestern United States is caused by *Lygus hesperus* Knight (Clancy and Pierce, 1966), which comprises 95 percent of the lygus population in Arizona (Telford *et al.*, 1965). In the agro-ecosystem, the primary host of lygus is alfalfa, on which populations increase to large numbers during the summer (Stern, 1969). When alfalfa fields are cut or become dry, lygus adults migrate into adjoining crops; these crops include cotton, deciduous fruits, and alfalfa seed (Davis, 1976). Migration results in infestations requiring insecticidal treatments. These treatments also reduce populations of predators and parasites of lepidopterous insects and spider mite pests, (Stern and Mueller, 1968). Secondary pest problems caused by insecticidal control of lygus adults in cotton have accentuated the need for alternate lygus-control methods. The strip-cutting system reported by Stern *et al.* (1964) which provided two growth stages of alfalfa in the same field at all times, reduced lygus emigrations into cotton. It was also found to partially control lygus bugs in seed alfalfa (Stern *et al.*, 1967).

Rakickas and Watson (1974) compared seasonal population fluctuations of adult and immature lygus and seven predaceous groups, before and after cutting, in strip-cut alfalfa. In addition to the herding of lygus back and forth within the alfalfa field throughout the season, of particular importance was the maintenance of large predator populations for the entire season. Perkins and Watson (1972) reported on the importance of one of these predators on lygus.

In an experimental field of cotton adjacent to a strip-cut alfalfa field, lygus populations did not exceed the economic threshold until August 19, too late to cause economic damage to the crop (Watson, unpublished). Whenever alfalfa was cut, allowed to get too far past bloom, or when the plant condition degenerated, lygus bugs left alfalfa and entered cotton. When alfalfa was cut all at one time, there was a period of time when excessive lygus bug damage was done to cotton. If only part of the alfalfa was cut (usually as strips through each alfalfa area) the lygus bugs would disperse—some would go to cotton temporarily, but most settled on the uncut alfalfa. The total seasonal predator populations were considerably higher in strip-cut alfalfa than in solid-cut alfalfa. Many routine farm practices cause predator populations to decline sharply by mid-summer in Arizona.

A modification of the strip-cut system whereby 40-acre blocks of alfalfa were divided into alternately-cut 20-acre blocks has produced satisfactory results. Utilizing this system, no lygus control was needed on adjacent cotton for an entire season (Watson, unpublished).

Stern (1969) has also utilized the alfalfa strip-plot method to attract lygus and keep them out of cotton. In this method, strips of alfalfa were planted within large cotton fields to attract migrating lygus and prevent further movement into the cotton. Watson (unpublished) has employed modifications of this system, using smaller strips of alfalfa than did Stern. Although large numbers of lygus were trapped within the alfalfa strips, economic levels did develop in the adjacent cotton. There is no doubt that strip-cutting of alfalfa is a sound principle and, if properly employed, will prevent or reduce the needed for insecticidal control of lygus in adjacent cotton.

Crop diversification

Natural ecosystems have diverse flora and fauna. This condition creates opportunities for complex interactions among the components, including arthropods and their natural enemies. These interactions tend to produce stability with pest populations often below damage levels. In farming areas pest problems are often less severe in regions of diverse agriculture than in regions tending toward monoculture. Strip planting of alfalfa in cotton (Rakickas and Watson, 1974) or planting alfalfa in the vicinity of cotton fields, and planting grain sorghum near cotton (Fye and Carranza, 1972) are examples of planned diversity. Diversity resulting from the "dirty field technique" (allowing noncrop plants to survive, rather than using clean culture) has a similar stabilizing effect on the agroecosystem. Diversification must include the proper plants, or the affects may be detrimental. Producing diversity through regionally-planned cropping sequences may be a viable approach for the future, but can be in conflict with economical crop production and equipment use.

Elimination of alternate hosts

Many insects, especially aphids, develop on alternate host plants in the vicinity of susceptible crops. Removal of those hosts delays or prevents development of pest populations. A management project against the green peach aphid, *Myzus persicae* (Sulzer), in southeastern Idaho is largely dependent upon removal or occasionally insecticide treatment of peach trees, the alternate winter host. The program also includes legislative procedures to prevent importation and/or distribution of aphid infested bedding plants. Over a 4-year period green peach aphid numbers on over 200,000 acres of potatoes in the region have been reduced to insignificant levels (Sandvol *et al.,* unpublished). Growers are gradually discontinuing preventive insecticide treatments, which were previously routinely used to prevent introduction and spread of potato leaf roll virus transmitted by the green peach aphid.

Tillage and Other Short-Term Operations

These are the practices that can be considered on a seasonal basis, often on individual farms. The over-all effectiveness is generally greater, however, if area-wide use is made of these pest-reducing techniques.

Cultivation

Cultivation affects a variety of pests that have at least part of their development in the soil. This includes cutworms, white grubs and grasshoppers. Effectiveness of cultivation often depends upon soil types and local conditions. General application is also limited because operations that reduce numbers of one pest may benefit other pests. Crop rotation and cultivation are the most widely used cultural practices for pest control, although growers do not always recognize these benefits.

Burning or flaming

Burning or flaming to destroy crop residues that harbor pests or disease organisms has been practiced on grass seed crops and cereals, but recently there has been considerable opposition to use of these measures because of smoke pollution.

In the eastern United States, workers found that flaming or burning could be used to destroy alfalfa weevil adults and eggs in late winter with little or no damage to the alfalfa. Field-flaming was studied under project W-84. These studies (Hansen and Simpson, 1969; Idaho and Utah, unpublished) demonstrated that stubble burning of alfalfa fields did not provide adequate alfalfa weevil control. Flaming did not control weevils in western United States primarily due to the relationship of the weevil development to the seasonal growth of alfalfa. Adult activity and egg deposition does not occur during periods of alfalfa dormancy in the West. If burning was delayed until the weevil activity started, the alfalfa was damaged. In Colorado, burning also destroyed exposed cocoons of the parasite *Bathplectes curculionis* (Thomson).

Destruction of crop residues

Many pests remain in seeds, stalks or other crop residues after harvest. Removal or incorporation of these residues into the soil by tillage is essential in the control of certain species like the European corn borer, *Pyrausta nubilalis* (Hubner). Destruction of chaff and screenings following the harvest of alfalfa and clover seed aids in the control of the seed chalcids, *Bruchophagus* spp.

Irrigation

It has long been recognized that excessive moisture is detrimental to spider mites. Water has been used commonly to suppress mites on house plants. Recent experiments in British Columbia and Washington showed that populations of the McDaniel spider mite, *Tetranychus mcdanieli* McGregor, were lower and their development slower in apple orchards where over-tree sprinklers were used, compared to undertree sprinklers (Hudson and Bierne, 1970; Hoyt, unpublished). The effect was probably physical but alteration of the microclimate could also be involved. The results of these studies fell short of expectations as no additional control could be measured in orchards with adequate predators.

Two common methods of irrigating alfalfa are flood irrigation and sprinkler irrigation. Following intensive alfalfa weevil sampling, it was noted that several of the sprinkler-irrigated fields had fewer weevil larvae and less damage than adjacent flood irrigated fields. Subsequent work showed that there was an increased mortality of instar I weevil larvae associated with sprinkler irrigation. This mortality is roughly proportional to the duration of the irrigation during the period of egg hatch and early instar I development. There may be several causes of this mortality, but the only one which was verified was the moisture barrier preventing newly hatched larvae from reaching the growing tips (Davis, unpublished).

Form of harvesting

Pest numbers on some crops, especially forages, can be reduced by clipping or early harvest to destroy immature insects that are in or on the foliage. This practice may also reduce damage by putting plant growth out of phase with pest development. Since 1965 many growers have switched from conventional cutting and baling to a system of green chopping. The alfalfa is cut and chopped about 10 days earlier than with the traditional harvesting system, then either fed directly to livestock, dehydrated, pelleted, or put into silage. This system was evaluated for alfalfa weevil control (Davis 1976). An exact percentage control could not be determined, but based on total weevil populations during the season and on expected weevil damage, the control was in the vicinity of 60 per cent. It prevented economically significant damage in many fields. Weevil damage was virtually absent in Idaho fields harvested for alfalfa pelleting (Bishop, Idaho, unpublished).

Crop-free periods

Certain soil-inhabiting pests including cutworms and wireworms can be controlled by crop-free periods combined with clean cultivation. This practice may not be economically attractive unless the problem is unusually severe or where it can be used in conjunction with fallowing to conserve moisture and control weeds.

Weed control

Weeds are important in maintaining pests before and after the crop-growing season and often serve as principal buildup hosts. Therefore, suppression of certain species of weeds within or near crops frequently benefits pest-control programs. Populations of the beet leafhopper, *Circulifer tennellus* (Baker), increase on Russian thistle and other desert plants and then invade crop lands. Weeds also serve as a source of the curly top virus which the leafhopper transmits to beets, beans and other vegetables (Carter, 1930). Spider mite problems are less severe in orchards where broadleaf weed hosts are controlled or sod cover crops are used (Davis and Nielsen, 1958). Various weeds have been associated with infestations of the green peach aphid in Idaho seed potato fields (Bishop, 1965), and weeds in fruit orchards are important sources of green peach aphids that infest various crops in the Columbia River Basin in Washington (Tamaki, 1975).

In contrast to the detrimental role of some weeds, others may benefit pest control. Several species of adult parasites require nectar in order to survive long enough to parasitize their hosts effectively. The tachinid, *Lixophaga sphenophori* (Villeneuve), attacking the New Guinea sugarcane weevil, *Rhabdoscelus obscurus* (Boisduval), in Hawaii was shown to require nectar which was usually obtained from the common garden spurge (*Euphorbia hirta*). The white grub parasites, *Campsomerus marginella modesta* Smith (Scoliidae) and *Tiphia segregata* Crawford (Tiphiidae), also required nectar. These latter parasites used nectar from several different plants (Beardsley, unpublished).

A similar requirement to that of the parasites exists with several predators. Many ladybird beetles and lacewings require vegetation for winter protection, and an early season food supply of either nectar or some host insect. This dietary requirement is the basis for the food supplement studies discussed in the previous paper.

Common practices in intensive agriculture usually include a nearly complete removal of noncrop plants in and around fields. If more effective use of beneficial parasites, predators and pollinators is to be accomplished, a reexamination of weed control practices is necessary. Strategic placement of important nectar plants in agricultural areas clearly improves the effectiveness of several parasites and predators. The noncrop plants required by each species of beneficial insects are not always the same. This means that a different pattern of weed control or planting of supplemental host plants will be needed for each crop system. At present only a few of these necessary food plants have been identified, but without question the list will be expanded in the future.

Brush drags or harrowing

From 1920 to 1940 cultural control of the alfalfa weevil was used extensively. The primary method was to harrow lightly or use a drag made of chicken wire and brush. This was done soon after the first crop harvest. When organic insecticides became available, most growers abandoned these cultural practices. About 1962 the alfalfa weevil became resistant to chlorinated insecticides, and the problem of pesticide residues on forage crops was recognized. Some of the older control recommendations were re-evaluated. Harrowing was compared to insecticide treatments applied to the stubble immediately following the first crop harvest (Davis, unpublished). Control was determined based on alfalfa regrowth plus weevil larvae surviving to feed on the new growth. About 40 percent control was achieved by harrowing, while several chemicals gave better than 80 percent control. The level of control was, by itself, not adequate. However, if combined with other cultural practices economic control levels can often be attained.

Multifacet use of Cultural Practices

The pink bollworm, *Pectinophora gossypiella* (Saunders), is a key pest of cotton in western Arizona and southern California. Pest management practices against this pest are interrelated to control practices on all crops in the area. Since its spread into this area in 1965 (Watson and Fullerton, 1969), multiple applications of insecticides have been used which have destroyed many predators and parasites. Notable among new problems which developed was the cotton leafperforator, *Bucculatrix thurberiella* (Busck), (Watson and Johnson, 1974). Hagen *et al.* (1971) discussed this problem and attributed these outbreaks to the destruction of nine species of parasites and ten of predators.

Cultural control programs were developed in Texas to control the pink bollworm (Adkisson and Gaines, 1960; Noble, 1969). Adkisson and Gains (1960) listed the primary practices as follows: (1) defoliate or desiccate the mature crop to cause all bolls to open at nearly the same time, thus expediting machine harvesting; (2) harvest the crop as early and in as short a time as possible; (3) shred stalks immediately after harvest: (4) plow stalks under immediately after shredding, thus preventing regrowth of new fruiting forms that might provide food for diapausing larvae; (5) prepare land for planting of the subsequent crop, including pre-plant irrigation in arid areas: and, (6) plant new crops during a designated planting period that allows for maximum "suicidal-emergency" of overwintering moths, (i.e., moths that emerge and die before cotton fruit is available to them for oviposition). The combination of practices developed in Texas provided the growers with a control program for the pink bollworm so successful that insecticides are seldom needed (Adkisson, 1972).

In Arizona, data have been accumulated showing the effects of various cultural practices on winter survival and spring moth emergence (Watson and Larsen, 1968; Watson *et al.*, 1970; Slosser and Watson, 1972). These data have shown that with each additional cultural control practice imposed on cotton (discing following shredding, plowing following discing, irrigation following plowing, etc.) there was an additional reduction in moth emergence. With the long growing season and mild winter conditions of Arizona, over-winter survival and spring emergence were still too great, regardless of the type of cultural practices used, to prevent economic infestations during the next growing season. The best results in the earlier cultural control investigations were achieved by shortening the growing season by earlier plowdown dates. These results showed that progressively higher moth populations emerged the following spring with each delay in plowdown date, ranging from September 1 to January 5 (Watson *et al.*, 1974).

These results prompted a 3-year study of pink bollworm control in relation to cotton management practices, particularly early crop maturity. The two major variables in the study were pink bollworm population levels during the growing season and irrigation cut-off dates, thereby affecting crop maturity dates.

The incidence of diapause in the pink bollworm is low until the last week of September, at which time it increases rapidly. Therefore, cotton where most green bolls are at least 3 weeks old by the last week of September should be incapable of building large overwintering pink bollworm populations. Thus a primary objective in shortening the growing season was to prevent large overwintering populations and thereby eliminate or reduce the need of insecticidal control of the pink bollworm during much of the next growing season. This would enhance pest management on cotton as well as on other crops in the area.

In this study the large plots consisted of four irrigation termination dates varying from mid-July, to early September, which corresponds to general grower practices. Subplot treatments within each irrigation termination date consisted of three levels of pink bollworm populations regulated by insecticides. These treatments were: (1) untreated check; (2) complete boll protection by initiating scheduled sprays at approximately 6-day intervals, starting at first boll formation; and, (3) a spray schedule initiated at 15 percent boll infestation. Results differed slightly from year to year, but the general trend was the same. The July 15 termination date was too early because of significant yield reductions. However, the latter three dates resulted in comparable yields within each of the three years.

The pink bollworm caused significant yield losses in all irrigation termination plots if left untreated. When pink bollworm yield data from all termination dates were combined, significantly greater yields were obtained when insecticide treatments were started at first boll formation rather than waiting until 15 percent of the bolls were infested. However, when examining yields relative to each termination date individually, no significant differences were shown between the plots where sprays were initiated at the 0 percent or 15 percent boll infestation level. The sprays started at the 0 percent level resulted in nonsignificantly greater yields at each date. These results are similar to those found in other studies (Watson and Fullerton, 1969). Except for the first irrigation termination date, July 15, the untreated check resulted in significantly less cotton than did either of the insecticide schedules.

Another interesting aspect of this study involved spring moth emergence relative to the previous seasons' cultural and chemical control practices. If only the untreated check plots are viewed in relation to termination dates and spring moth emergence it was found that for each subsequent cut-off date greater numbers of moths emerged the following spring.

Delaying cotton planting dates so that most of the spring emergence of pink bollworm adults has taken place before fruiting forms are available has been proposed as another alternative for control by cultural means. This plan does not seem to be as acceptable in Arizona because considerable yield reductions occur with delayed plantings. Additionally, it would be necessary to maintain the cotton as late in the season as possible, at which time large diapause populations would develop.

To summarize, it appears that management of irrigation water in Arizona and southern California to effect earlier than normal termination of cotton production would improve pest management and biological control of pests on cotton as well as on other nearby crops. This can be timed so no yield reduction is experienced, and at the same time it prevents build-up of large overwintering pink bollworm populations. On an area-wide basis and with better in-season control, the severity of the problem should decrease. With fewer insecticide applications needed for in-season control of the pink bollworm, plus the reduction of lygus bug numbers through alfalfa management close to cotton fields, naturally-occurring biological control of a number of secondary pests is improved.

Resistant Crops

Use of resistant varieties is one of the most convenient and effective methods of pest control. Implementation requires only a change in seed or planting stock. Crop resistance has been of primary importance in control of pests on cereals and other field crops and on certain horticultural crops.

Development of resistance has been directed toward complete or near complete control of target pests. Recently, however lower resistance has been recognized as important in reducing insecticide usage on crops where primary controls are chemical. Partial resistance has proved valuable as a component in integrated control programs on alfalfa, cotton,and other crops.

Initial costs of developing resistance may be high since considerable time and effort may be required for its incorporation into varieties otherwise horticulturally or agronomically acceptable. But resistant varieties pay high returns because they can provide long-term solutions to pest problems.

Insects have produced biotypes that overcome resistance when the resistance in the plant is controlled by one or a few genes (vertical resistance). This has happened with the Hessian fly (Gallun, 1974) and the spotted alfalfa aphid, *Therioaphis maculata* (Buckton), (Nielson, *et al.* 1971), but in each instance varieties resistant to the biotypes were developed.

Crop resistance is not universally applicable because usable levels of resistance against a particular pest may not be present in the gene pool of the plant, or resistance may be linked with undesirable plant characteristics. Plant resistance to grasshoppers, cutworms or other pests without specific host plant relationships is not common.

Discussion

Cultural control practices, like biological controls, are population-lowering approaches. Therefore complete or even high levels of control are rarely achieved through this method alone. Pest numbers can be reduced, however, so that the amount of pesticide required is much less. The basic decision whether to attempt pest control by cultural methods is dependent on economic injury levels in that particular crop.

Cultural methods have an important place in the control of pests. These methods are economical, easy to apply, environmentally safe, and readily integrated with other types of control methods. However, their importance is often overlooked because effects are often indirect and subtle, and they often do not give adequate control independently. As the principles and goals of these methods become more widely understood, appreciation and support should follow.

Because cultural controls are largely preventive, effectiveness is difficult to demonstrate. Growers hesitate to place full confidence in them until they understand the principles involved. Such understanding will also help growers appreciate why exact timing and following precise procedures is required in implementation. Effective use of cultural controls also requires that growers think in terms of long-range control rather than for convenience and short-term profits only.

Cultural practices fall into several categories, but much attention is being given to cultural practices which favor natural enemies or increase the impact of physical factors in the environment. It is safe to say that there are cultural manipulations which can reduce pest problems in any crop. From the growers' viewpoint, the big question is one of economics: Can a greater return from an investment come from changing a cultural practice or from using a pesticide? From technical and long-range viewpoints, however, the prevention of potential resistance to pesticides may be most important.

It is much harder to evaluate cultural practices, and to prove their effectiveness in pest control, than it is with chemical pesticides. A randomized experiment can be conducted using a pesticide, then control evaluated at given intervals by comparisons to an untreated check. This is not always possible with cultural practices. Many cultural activities, such as destruction of crop residues, weed control, etc. have their effects on an area basis. Replicated plots in a single field would prove nothing. Selling growers on the necessity to modify cultural practices without direct and positive proof of effectiveness is difficult. The best evidence for effective control through cultural manipulation must come from a series of large-scale demonstrations. In fact, only when isolated farming communities, or concentrated units of some particular crop, practice the same procedure can the full impact be observed.

A serious difficulty with modifying cultural practices is the complexity of any agronomic change. A practice which is used to control insects may not control weeds. The timing of a particular activity needed to control a pest may not be compatible with planting or harvesting dates. Planting of crops so that they are the same age over an entire area will often control insects, but it is usually impossible from a management viewpoint. Whenever pest control through cultural manipulation is contemplated, it is extremely important to consider all ramifications of crop production and economics.

An example of this complexity is the earlier harvest of first-crop alfalfa in northern Utah to control the alfalfa weevil. Cutting at the bud stage rather than 10 percent bloom will give about 60 percent control of the weevil. The immediate questions arising from this practice include the following:

1. Is 60 percent control adequate and won't pesticides be needed anyway?
2. Rainy periods often occur during late May and early June. If I plan to cut early, then cannot do so, how do I control the weevil?
3. Earlier cutting reduces total weight of the hay, but increases protein. Am I better off economically?
4. Will earlier cutting increase the danger of bloat in cattle?
5. What happens to the longevity of the alfalfa stand when it is cut early?
6. What happens to the weed problem after early cutting?
7. What does early cutting do to other insects in the field?
8. Early cutting requires a change in irrigation practices. What effects will this have?
9. Should each hay crop be cut earlier or only the first crop?
10. What is the long-term effect of relying totally on pesticides for control?

A similar series of questions can be asked regarding almost any change in agronomic practices. Many cotton growers in Arizona are not convinced that early harvest followed by destruction of the crop residue will result in enough savings from reduced amount of insecticides to offset possible loss in yield. The fruit growers in Washington are still apprehensive over the possibility of increased disease problems with overhead sprinkler irrigation. Sugar cane growers in Hawaii are not ready to accept the weeds necessary to support parasites. Cotton growers in California must be educated to accept fence-row vegetation necessary for green lacewing build-up.

Application of cultural controls may produce conflicts of interest. Actions taken against one pest may benefit another. Leaving weeds to enhance biological control in one crop may create problems in other crops or for other growers. Those conflicts involving single farms can be worked out in favor of the total operation, but inter-farm conflicts are more difficult to solve and may limit applications of certain methods.

The importance of cultural practices in modern pest control has been demonstrated in several cropping systems, but most strikingly in management of cotton insects where a variety of crop management and tillage practices were combined with biological and chemical procedures to produce a successful integrated system. These demonstrations of the flexibility of cultural methods, added to their advantages from an environmental standpoint, will provide incentives for broadening present uses and for developing new methods tailored to developing integrated programs. Systems based upon cooperation should become more common, utilizing area coordination of planting sequences and planting and harvesting dates, and carefully planned schemes for plant diversity. But adoption of area programs will depend upon growers' williness to risk short-term profits for the sake of long-range benefits.

Resistance to pests and the selection of proper cultivars is difficult to include in a discussion of this type. In part, it is cultural control of pests, but it could also be considered as biological control or even a category by itself.

Not only must the questions relating to changed cultural practices be answered from many angles, but a diligent effort is needed to educate growers to change cropping practices. This is especially true when scientists reverse their long-standing ideas. For example, for at least 100 years, housefly control centered on complete manure removal and sanitation. Now it is found that a significant level of biological control of houseflies occurs when some manure is allowed to remain to harbor the predators and parasites.

Even more involved long-range policy questions are being raised. For economical and production reasons we have developed vast regions of monocultures or limited crop diversity. However, fewer pests occur in diversified agriculture. What sort of compromise is needed to achieve a balance between these opposite philosophies?

Regardless of the complexities of changing practices in agricultural production (and clearly demonstrating the advantages to the growers) there is no doubt that pest control through cultural methods will always be important. Often in the past, insect control was almost a side benefit of agronomic practices. Fall plowing, crop rotation, etc. are valuable insect-control techniques, but growers usually plow to prepare the soil for planting, and rotate crops to improve soil fertility. Of the common agricultural pests only weeds are normally considered directly by farmers in their cultural practices. Actually, however, insects, diseases and nematodes are also susceptible to these approaches. An attempt must be made to establish the "optimum" cultural practices from the standpoint of yield, economics, and pest population regulation.

If pest management is to work, ever-increasing attention must be devoted to cultural control practices. The entire concept is built on the precept that population-lowering techniques must form the foundation. The more direct control methods, such as pesticides, must be superimposed on a biological and cultural control foundation. Before sound cultural control methods can be devised, a detailed knowledge of the pest and a thorough understanding of the crop itself are necessary. Each pest, on each crop and in each area, may require modifications in the cultural control approach.

Literature Cited

ADKISSON, P. L.
1972. Use of cultural practices in insect pest management. *In:* Implementing Practical Pest Management Strategies. Proc. Nat. Ext. Insect Pest Management Workshop. Purdue Univ., Lafayette, Ind. March 14-16,1972. Pp. 37-50.

ADKISSON, P. L. and J. E. GAINES.
1960. Pink bollworm control as related to the total cotton insect program of central Texas. Tex. Agric. Expt. Sta. Misc. Publ. 444. 7 pp.

BISHOP, G. W.
1965. Green peach aphid distribution and potato leafroll virus occurrence in the seed-potato producing areas of Idaho. J. Econ. Entomol. 58(1): 150-53.

CARTER, W.
1930. Ecological studies of the beet leaf hopper. U.S. Dept. Agr. Tech. Bull. 296: 1-115.

CLANCY, D. W. and H. D. PIERCE.
1966. Natural enemies of some lygus bugs. J. Econ. Entomol 59(4): 853-58.

DAVIS, D.W. (editor).
1976. Insects and nematodes associated with alfalfa in Utah. Utah Exp. Sta. Bul. 494. 59 pp.

DAVIS, D. W. and G. L. NIELSEN.
1958. The biology and distribution of the McDaniel mite in Utah. Utah Acad. Sci. 35: 166.

GALLUN, R. L.
1974. Genetic interrelationships between host and organism and influence on resistance. pp. 395-402. *In:* Maxwell, F.G. and F.A. Harris (editors), Proceedings of the Summer Institute on Biological Control of Plant Insects and Diseases, Univ. Press of Mississippi, 647 pp.

HAGEN, K. S., R. VAN DEN BOSCH, and D. L. DAHLSTEN.
1971. The importance of naturally-occurring biological control in the western United States. pp. 253-93. *In:* Huffaker, C.B. (editor), Biological Control, Plenum Press, N.Y., N.Y., 511 pp.

HANSEN, R. W. and R. G. SIMPSON.
1969. Agricultural flaming for insect and weed control in Colorado. Colo. Exp. Sta. Bul. 538 2: 1-9.

HUDSON, W. B., and B. P. BEIRNE.
1970. Effect of sprinkler irrigation on McDaniel and European red mites in apple orchards. J. Entomol. Soc. Brit. Columbia 67: 8-13.

NIELSON, M. W., M. H. SCHONHORST, H. DON, W. F. LEHMAN, and V. L. MARBLE.
1971. Resistance in alfalfa to four biotypes of spotted alfalfa aphid. J. Econ. Entomol. 64(2): 506-10.

NOBLE, L. W.
1969. Fifty years of research on the pink bollworm in the United States. USDA. Agr. Handbook No. 357 62 pp.

PERKINS, P. V. and T. F. WATSON.
1972. *Nabis alternatus* as a predator of *Lygus hersperus.* Ann. Entomol. Soc. Amer. 65(3): 625-29.

RAKICKAS, , R. J. and T. F. WATSON.
1974. Population trends of *Lygus* spp. and selected predators in strip-cut alfalfa. Environ. Entomol. 3: (5): 781-4.

SCHLINGER, E. I. and E. J. DIETRICK.
1960. Biological control of insect pests aided by strip-farming alfalfa in experimental program. Calif. Agr. 14(1): 8-9, 15.

SLOSSER, J. E. and T. F. WATSON.
1972. Influences of irrigation on over-winter survival of the pink bollworm. Environ. Entomol. 1: 572-76.

STERN, V. M.
1969. Interplanting alfalfa in cotton to control lygus bugs and other insect pests. Proc. Tall Timbers Conf. Ecol. Anim. Control Habitat Mgm't. 1: 55-69.

STERN, V. M. and A. MUELLER.
1968. Techniques of marking millions of insects with micronized fluorescent dust with especial emphasis on marking millions of *Lygus herperus* for dispersal studies. J. Econ. Entomol. 61: 1232-7.

STERN, V. M., R. VAN DEN BOSCH and T. F. LEIGH.
1964. Strip-cutting alfalfa for lygus bug control. Calif. Agric. 18(4): 4-6.

STERN, V. M., R. VAN DEN BOSCH, T. F. LEIGH, O. D. MCCUTCHEAN, W. R. SALLEE, C. E. HOUSTON, and M. J. GERBER.
1967. Lygus control by strip cutting alfalfa. Univ. Calif. Agric. Ext. Serv. AXT-241. 13 pp (out of print).

TAMAKI, G.
1975. Weeds in orchards as important alternate sources of green peach aphids in late spring. Environ. Entomol. 4: 958-60.

WATSON, T. F., K. K. BARNES, J. E. SLOSSER, and D. G. FULLERTON.
1974. Influence of plowdown dates and cultural practices on spring moth emergence of the pink bollworm. J. Econ. Entomol. 67: 207-10.

WATSON, T. F. and D. G. FULLERTON.
1969. Timing of insecticidal applications for control of the pink bollworm. J. Econ. Entomol. 62: 682-85.

WATSON, T. F. and P. H. JOHNSON.
1974. Life cycle of the cotton leaf perforator. Prog. Agric. in Ariz. 24: 12-13.

WATSON, T. F. and W. E. LARSON.
1968. Effects of winter cultural practices on the pink bollworm in Arizona. J. Econ. Entomol. 61: 1041-44.

WATSON, T. F., W. E. LARSON, K. K. BARNES and D. G. FULLERTON.
1970. Value of stalk shredders in pink bollworm control. J. Econ. Entomol. 53 : 1326-28.

Chapter 9.
SELECTIVE PESTICIDES

D. W. Davis and S. C. Hoyt

Introduction

Contrary to the thinking of some individuals, and possibly the wishful thinking of others, pest management does not exclude the use of pesticides. According to Smith (1970): "Pesticides remain, in spite of their adverse publicity, our most powerful tool in the management of pests. The pesticides, as developed in the last 30 years, are highly effective and economical and can be marshalled quickly to have an immediate impact on the pest population—even over a large area. When pest populations approach economic levels, there is little other than pesticides we can use to avoid damage and which will have the desired immediate effect. Hence, it seems clear that pesticides must and will continue to be used in a major way in pest management. In fact, few informed and rational people desire or anticipate the elimination of all pesticides."

The really critical question is how to use pesticides so that their maximum benefits can be obtained while adverse effects are minimized. From the viewpoint of biological control and pest management this means preserving beneficial insects such as predators, parasites and pollinators while at the same time maintaining numbers of harmful pests below their economic injury levels. In order to accomplish this we must think in terms of selective pesticides and/or selective use of pesticides.

All pesticides are selective to a certain degree. For example, a weed-killer, or herbicide, is selective in that it kills weeds but rarely kills insects. A truly nonselective pesticide would be one that kills all living organisms either beneficial or harmful without any consideration as to the type of living organism involved. Therefore, the term 'selective pesticide' is a relative term. Just how selective do you want a material to be? A totally selective material would be one that would kill only one target species of organism and nothing else. The only controls known at present which approach this make use of insect viruses, specific parasites, or pheromones. If a company were to develop a totally selective pesticide, it would have to go through the entire expense and research procedure required for the broad-spectrum materials. It is questionable whether this could be done economically, as it costs approximately 10 million dollars to develop a new pesticide. A pesticide which controls only one pest species would have to be used against a pest of great economic significance in order to reach a large-enough sales volume to regain this investment. Few pest species are this important. Most scientists do not visualize 100 percent selectivity but, rather, desire sufficient selectivity to protect the major beneficial organisms in the ecosystem.

There are two main types of selectivity. First, due to its inherent toxicity a pesticide may kill only certain types of organisms—for example, an acaricide may kill mites but not insects. Some acaricides may be even more specific and control spider mites, while sparing certain predatory mites. The second type of selectivity is associated with use patterns—for example, some short-lived insecticides can be used in the evening and not leave enough chemical on the crop to kill pollinators the following morning. Another example can be based on seasonal timing—if a particular pesticide is applied early enough in the season to avoid the beneficial insects, it achieves selectivity. A third example would be careful timing of a pesticide so that minimum rates could be used. These minimum rates may be effective against a pest species but at the same time may allow a large proportion of the desired organisms to survive.

To use pesticides within the context of a pest management system, a change in thinking must occur. The traditional idea is that the control of a pest species must depend entirely upon the pesticide. This means that about a 98 percent control of that pest is attempted. When using selective pesticides in a pest management system this high degree of control is often unnecessary. If sufficient pests are killed to achieve an immediate benefit, the remaining individuals will serve as food for predators and parasites. This allows the beneficial insects to finish the job started by pesticides, and it also results in reduced selection pressure for resistance of the pest to the pesticide.

When a pesticide application results in a nearly complete control, there is often a resurgence of pests due to the absence of predators and parasites. This resurgence may be of the original target pest or some previously minor pest. The grower then becomes committed to a sequence of drastic chemical actions. This sometimes has resulted in 15 or more pesticide applications on one crop in a given season with a high pressure for the selection of resistant insects. In contrast, when predators and parasites are allowed to survive the resurgence of pest populations rarely occurs and selection pressure is reduced. Therefore, control can be accomplished with fewer and less potent treatments. It becomes necessary to educate growers to accept a limited number of pests early in the season to avoid large numbers later. Learning to accept limited pest numbers will be difficult for many individuals. It is necessary to understand that insects become pests only when their numbers exceed certain levels. A single individual insect or weed in itself is not serious. A successful pest management program requires acceptance of low-level pest populations which will be kept below economic injury levels through occasional applications of selective pesticides or other regulating mechanisms.

The use of selective pesticides cannot be isolated from other aspects of pest management. A greater knowledge of the biology and population dynamics of both pests and natural enemies in the ecosystem is needed than with traditional insecticidal controls. This must be coupled with a knowledge of the economic impact of the pest on the crop. Good sampling techniques must be available to take maximum advantage of selective use patterns. Selective treatments should be available for all pests, as a single non-selective application can upset the management program and lead to pest outbreaks.

Selectivity Based on Nature of the Pesticide

Many early insecticides were inherently selective. The arsenicals, for example, killed those insects with chewing mouthparts. Predators, parasites and insects with piercing-sucking mouthparts such as aphids and leafhoppers were not killed. Nicotine sulfate was effective against insects with soft bodies, while those with hard bodies usually escaped. With the advent of the synthetic organic insecticides, new problems arose. As these materials kill insects through various combinations of contact, ingestion, and fumigant action, they are generally less selective, and predators, parasites and pollinators are more seriously affected. Even with newer insecticides, however, some selectivity exists. Systemic materials such as demeton (Systox) can be used to control harmful insects while preserving many of the beneficial insects. Once demeton is absorbed into the foliage of a plant, it no longer kills by contact action. Some of the short-lived materials such as TEPP are highly toxic to the active insect stage at the time of application. However, insect eggs, pupae and transients such as pollinators not in the field during spraying, are frequently unaffected. When the chlorinated insecticides were first introducted they were considered to be effective against virtually all types of insects. It was found, however, that DDT, toxaphene, and several of the other chlorinated insecticides, are quite selective. Toxaphene, for example, is not highly dangerous to honeybees. Several of the organophosphates can also be selective. Trichlorfon (Dylox) can be used to control lygus bugs and caterpillars with relatively little harmful effect on many predators and pollinators.

In developing new pesticides, the companies formerly tested for action against a relatively few pests. As the concept of pest management has developed they have begun looking for selectivity. This has resulted in revised screening processes and rechecking of many older materials.

When the new synthetic organic insecticides became available, most research entomologists looked for effects on pests only. The preservation of beneficial insects, however, became more and more important as clearer concepts of the agroecosystem developed. At present most research with insecticides considers effects on a broad range of both pest and beneficial insects. This concept of pesticide use and evaluation has been fostered by regional project W-84, centering primarily on alfalfa, cotton and apples.

Alfalfa insect studies

Alfalfa insects, particularly the alfalfa weevil, have been studied in Colorado, Idaho and Utah. In all three states, insecticides have been included in the studies. Six different insecticides are commonly recommended; however, their impact on the beneficial insects differs greatly. Carbofuran (Furadan) has been consistently the most effective material for control of the alfalfa weevil. Unfortunately, carbofuran also kills most other insects present at time of application. Also, it is only partially effective for aphid control but kills most aphid predators, and greater numbers of pea aphids and spotted alfalfa aphids commonly occur after its use. Carbofuran depresses populations of *Bathyplectes curculionis*, the alfalfa weevil parasite, up to two or three weeks and is extremely damaging to pollinators in alfalfa seed production. The primary way of using carbofuran in pest management programs is to time it so that applications cause minimum harm to beneficials. In Utah this is very early in the spring when alfalfa has first started to grow, or to the stubble following the harvest of the first hay crop.

Two other products which have given reasonably effective control of the alfalfa weevil are phosmet (Imidan) and methidathion (Supracide). Neither of these is as effective against the weevil as carbofuran, but they are less toxic to several predators and parasites. Therefore, applications made to control alfalfa weevils during the time of extensive beneficial insect activity makes them more acceptable. Unfortunately, neither is selective enough to adequately protect pollinators. Some insecticides such as methyl parathion kill most of the beneficial insects present in the field at the time of application, but the relatively short residual life allows for their rapid return. Insects in either the egg or pupal stage survive treatments with short-lived insecticides. In alfalfa, we can use chemicals with an inherent selective action or those with shorter residual life to make better use of biological control.

Alfalfa-seed producers, not interested primarily in alfalfa weevil control, can also use insecticides with selective properties. In the bloom period, lygus bugs and aphids frequently become problems in alfalfa-seed fields. The insect control program must be severely curtailed because of danger to pollinators. A few insecticides can be used during this period. Trichlorfon controls caterpillars, grasshoppers and lygus bugs while doing minimum harm to many of the predators and parasites, and it is virtually nontoxic to honeybees. Demeton controls aphids, thrips and lygus bugs and can be used in combination with trichlorfon. When applied in the evening demeton is taken into the plant tissue by morning, thus protecting honeybees and other pollinators.

Deciduous fruit insect studies

A successful integrated control program has been developed on apples and other deciduous fruits based primarily on work conducted in the state of Washington, but with contributions from several other states and Canadian provinces. The successes of this program depend in part on selective chemicals and selective use patterns. Carbaryl (Sevin), for example, is effective for the control of codling moth, but it is not used commonly as an insecticide because of its adverse effect on mite predators. However, carbaryl is used as a thinning agent on apples because selectivity can be obtained by restricting the application to the upper and peripheral parts of the trees. This allows the desired thinning yet conserves a source of predators for later spread to the entire tree. Phosmet, phosalone (Zolone), or reduced rates of azinphosmethyl (Guthion) are also effective against the codling moth but have considerably less impact on beneficial species than carbaryl. Selective aphicides, acaricides, and fungicides have also been identified so that a complete program which conserves beneficial species can be employed. In addition to conserving predators, the program is geared to avoid excessive destruction of the apple rust mite, an alternate source of food for the principal mite predators. This food source helps maintain adequate numbers of predators during periods when spider mite populations are low.

Cotton insect studies

In Arizona, cotton pest management studies have stressed insecticide selectivity through correct timing and placement. However, there appears to be considerable promise using inherently selective pesticides to control the pink boll worm. The potential for obtaining selective action with currently available insecticides appears to be greatest with insect pathogens.

Other studies not in W-84

While the preceding examples of pesticide selectivity were chosen from work being conducted by the members of regional project W-84, it should be pointed out that similar studies elsewhere are also showing successes with selective chemical control. Forest Service personnel, for example, are studying insect viruses and bacteria in the control of forest pests. When these have been used to control such pests as the tent caterpillar, the resultant impact on beneficial insects has been slight. In contrast, when they have used the most general materials such as carbaryl or DDT, there is usually a build up of minor pests due to the elimination of predators and parasites. The Forest Service has also tested mexacarbate (Zectran), which has shown considerable selectivity in control of the spruce budworm without causing excess damage to nontarget insects.

While the degree of selectivity will vary depending on the crop and pests involved, there are several materials which have aroused considerable interest in the last few years. Several of the newer acaracides have been selective in that they control the harmful spider mite with minimum damage to the beneficial mite predators. Noteworthy among these new acaracides have been cyhexatin (Plictran) and propargite (Omite). Some of the older acaracides, such as dicofol (Kelthane), are detrimental to predatory mites.

There has been continued interest in the systemic materials where selectivity is based on insect feeding habits. Unfortunately, several of the newer systemic materials also kill by contact action. Many of them do not achieve the same degree of selectivity as that obtained with demeton or oxydemetonmethyl (Metasystox-R).

Selectivity Based on Use Patterns

Because of the problems associated with commercial development of physiologically selective pesticides, the greatest potential for achieving selectivity at present appears to be in altering the use patterns of broad-spectrum pesticides.

Selectivity of some general pesticides can be obtained by such procedures as modifying dosages, timing, or application techniques. It should be pointed out, however, that altering use patterns does not insure that a pesticide will become favorably selective. A treatment may have desirable selectivity in one situation and be highly toxic to important natural enemies in another. Each material must fit the peculiar needs of the agroecosystem where it is to be used.

Dosage

The amount of a pesticide to use in a management program should be based on several factors. These include the degree of pest control required to prevent economic loss, the potential for control by natural enemies or cultural methods, the possible need to maintain a reservoir of pests as food for natural enemies, and the conservation of the natural enemies related to the entire pest complex on the crop. Although information may be lacking for some of these factors, it is important to recognize that a high level of control may be necessary for a few key pests only. The use of high dosages to overcome inadequacy of timing or coverage should be avoided.

Dosage and timing cannot be separated in many cases, because as dosages are reduced to achieve selectivity it is usually necessary to compensate for the reduced killing power with improved timing. This has been the case with reduced rates of azinphosmethyl in integrated programs on deciduous fruits in the West. Full dosage of this chemical (2 or more pounds active ingredient per acre) provides codling moth control for 4 to 5 weeks but is toxic to predacious mites. When the dosage is reduced to 0.75 to 1 pound a.i. per acre, the material is effective against codling moths for 21 to 25 days, predatory mites survive in moderate to high numbers, and predators and parasites of aphids return to the treated area in a short time. Further reduction in dosage is possible and can achieve even greater selectivity. Precise monitoring of codling moth populations and very careful timing of spray applications are necessary to prevent increased worm damage. At these extremely low dosages it is also more important to maintain periodic observations of the orchard for the presence of other pest species.

Timing of treatments

The importance of proper timing where reduced dosages are used has already been stressed. Timing can also be adjusted seasonally or daily to take maximum advantage of periods when beneficial species are not present on the crop, are inactive, or are in a more resistant stage. Examples of times to achieve selectivity were presented earlier for alfalfa. Compounds known to be moderately toxic to the predatory mite, *Amblyseius fallacis* (Garman), may be used selectively on apples in Michigan during April to mid-June as most of the predators are on the ground cover during that period (Croft and Brown, 1975).

Application techniques

Restricting the area of application of a chemical which is toxic to parasites or predators is another means of obtaining a selective effect. The most obvious method of accomplishing this is to carefully delineate the area of potentially damaging populations of a pest and treat only that portion of the field, allowing beneficial species to survive in, and migrate from the untreated portion. The use of carbaryl as a thinning agent on apples where the application is restricted to the upper and peripheral parts of the tree has been an important means of conserving predacious mites. Watson (in press) has reported on studies of a similar selective placement of azinphosmethyl on cotton. In this case the lower portion of the plant was treated to control pink bollworm, but the terminals, where *Orius* spp. are particularly numerous, remained untreated.

In Pennsylvania, a technique of spraying the middles of alternate rows of apple trees has been developed (Asquith, 1973). In this case spraying is done from the middle of even or odd numbered rows at one time, and 7 days later the other middles are treated. In contrast to blanket sprays applied at 14-day intervals, this procedure has allowed a reduction in dosage and provides better refuge for mobile natural enemies of pests.

Selectivity Based on Natural Enemies Becoming Resistant to Pesticides

A few species of natural enemies of pests have developed strains resistant to pesticides. Indigenous populations of these resistant natural enemies have gererally proved effective in pest management programs, and resistant strains of at least two species of predatory mites have been distributed to various parts of the world, colonized, and released in orchards. Several of the releases have resulted in successful integration of biological control of plant-feeding mites with chemical control of other pests. In some cases, release of these resistant predators has failed to improve the biological control of mites. This could be due to several factors, but the most important appears to be other toxic chemicals in the total program. The resistance is usually limited to a few related chemicals, and care must be taken in the selection and use of any chemicals to which the predators are not resistant. For example, several fungicides required for control of apple powdery mildew are toxic to predatory mites resistant to certain organophosphate insecticides.

Although most cases of resistance in natural enemies have been reported in predatory mites, a few species of insects have been shown to be resistant to pesticides, and it is quite possible that additional cases exist (Croft and Brown, 1975). Selected resistance generally is recognized in the pest species first, however, and the subsequent change to a different chemical for control may prevent the selection of even moderate levels of resistance in the natural enemy complex. While the possibility of artificially selecting resistant strains of parasites and predators presents an intriguing approach, the few attempts made so far have not been successful (Croft and Brown, 1975). Breeding of resistant natural enemies is not expected to play a major role in pest management programs in the near future.

Discussion

The typical agroecosystem consists of a single dominant crop species plus a few weeds. This monoculture may extend over large areas, usually with a limited diversity of organisms. The application of insecticides to this agroecosystem further reduces the numbers of certain insect species and leads to even more simplification. This lack of complexity often results in extreme fluctuations in populations of those species which persist. Any new factor entering into the environment can cause the system to react violently. This is typical of the pattern achieved with traditional insecticide control methods. It is on this premise that we try to use selective materials, or more general insecticides in a selective fashion, so that there is a minimum disturbance of this agroecosystem.

When a pesticide is introduced into a crop system, it should be used in a way that creates a minimum amount of damage to the nontarget organisms. From a farmer's point of view, however, the ecological significance is sometimes lost because he measures control by the immediate reduction in pest numbers, rather than by marketable produce. Furthermore, if a farmer is going to apply a material to control a multitude of pests, as he would with a more general pesticide, the cost would be based on one kind of chemical. If he uses selective chemicals, he may have to combine two, three, or even more, materials into his pest control program. This may increase the pesticide costs on a short-term basis. It is hard for a farmer to visualize future savings based on an increased reliance on biological or cultural practices.

Much of the added cost of each chemical will be offset by fewer repeated applications, lower dosages, and more carefully timed applications. It is highly probable that the cost of pesticides will increase. This is particularly true of those pesticides that are selective. The question has been raised in the minds of many individuals whether there should be a system within our government, or within other agricultural agencies to encourage or partially defray these added costs of developing selective pesticides. In other words, should a company decide to invest in the development of selective pesticides, a certain portion of the cost would be defrayed by the Government. The future of this approach is highly doubtful, as there has been no inclination toward cost subsidy by any govermental agency. Some specific control methods such as certain insect viruses have been developed directly by govermental agencies, but few chemical pesticides have come through these channels.

Fewer applications per season are necessary with chemicals which allow natural enemies to survive. With a pesticide which allows parasites and predators to survive and finish the job, there is little tendency for sudden increases of pest populations. Sudden increases in mite and scale populations were prevalent in the early days of DDT and parathion. The kill of the predators was often more complete than the control of the pests. About 3 weeks after application there often were higher populations of pests in treated than in untreated areas. With a selective material preserving the predators, this type of a flareup is prevented.

With a properly executed program using selective insecticides, applications are carefully timed by sampling techniques such as pheromone traps. The pesticide rate is less than that used in traditional pest control programs. There is much less tendency for nontarget pest species to switch from a minor role to a major one. This tendency for minor pests to become important has been noted especially in cotton production.

Concern has been expressed regarding the use of highly selective materials because of the complex of insect and mite species on a single crop. If pesticides are extremely selective, it may be necessary to use mixtures or sequential applications of several materials which will result in a loss of selectivity for the total pesticide program. The problem is further complicated if there are no selective chemicals available for one or more of the pests. If even one pest cannot be controlled selectively, the total program is not selective.

There is almost unanimous agreement that we must develop or identify more selective pesticides and learn to use them, but it is doubtful if many new ones will be developed in the near future. Watson (1975) states, "With the current difficulties in getting registrations for new pesticides, it will take several years before new truly selective pesticides can be developed for grower use." It appears, therefore, that the greatest opportunity lies in using the currently available insecticides in a manner more compatible with the total environment.

Selectivity exists with many of the experimental biological pesticides. However, the registration of certain of the insect viruses has been progressing slowly. There will undoubtedly be several of these materials available for commercial use in the future, but now only two bacterial insecticides and two insect viruses are registered for use.

Another class of selective pesticides being investigated is that of growth substances found naturally within the insect itself. These materials include insect growth regulators and juvenile hormones. Some of these chemicals are mimics of natural chemicals rather than the natural hormone itself. This approach to chemical control using modifications of natural materials within the insect body protects birds and mammals in an agroecosystem, but so far the selectivity toward protection of beneficial insects appears to be minimal.

Selective pesticides must be considered along with developments in other phases of insect control. We cannot isolate the chemical approach from biological or cultural control methods. There is a danger that partially-informed individuals will attempt to do everything with selective chemicals. We do not wish to make this implication. The use of selective chemicals while important in pest management, is only a portion of the management system. Consideration of all tactics must be incorporated if the pest management system is to function at maximum efficiency.

Selected References

ASQUITH, D.
1973. Some insecticides and miticides for integrated pest management in apple orchards — 1973. The Penn. Fruit News 52: 16-20.

CROFT, B. A., and A. W. A. BROWN.
1975. Responses of arthropod natural enemies to insecticides. Ann. Rev. Entomol. 20: 285-335.

GRASSER, R.
1966. Use of pesticides in selective manners. pp. 109-13. *In:* Proceedings of the FAO Symposium on Integrated Control Vol 2. 186 pp.

METCALF, R. L.
1974. Selective use of insecticides in pest management. pp. 190-203. *In:* Proceedings of the Summer Institute of Biological Control of Plant Diseases and Insects, ed. F. G. Maxwell and F. A. Harris, Univ. of Miss., 647 pp.

NEWSOM, L. D.
1966. Essential role of chemicals in crop protection. pp. 95-108. *In:* Proceedings of the FAO Symposium on Integrated Control. Vol. 2. 186 pp.

SMITH, R. F.
1970. Pesticides: their use and limitations in pest management. pp. 103-118. *In:* Concepts of Pest Management, ed. R. L. Rabb and F. E. Guthrie. North Carolina State Univ. 242 pp.

STERN, V. M., R. F. SMITH, R. VAN DEN BOSH, and K. S. HAGEN.
1959. The integrated control concept. Hilgardia 29(2): 81-101.

VAN DEN BOSCH, R. and V. M. STERN.
1962. The integration of chemical and biological control of arthropod pests. Ann. Rev. Entomol. 7: 367-86.

WATSON, T. F.
1975. Practical considerations in use of selective insecticides against major pests. pp. 47-65. *In:* Pesticide Selectivity, ed. J. C. Street. Marcel Dekker Inc., N. Y., N. Y. 197 pp.

Section III
IMPLEMENTATION OF PEST MANAGEMENT

Chapter 10.
PEST MANAGEMENT PROGRAMS

M. T. AliNiazee and E. R. Oatman

Introduction

The importance of regional components in pest management programs, and the limitations of excess extrapolation of results from one area to another, were discussed in the introduction of this publication. Basic concepts and approaches require a regional viewpoint while specific mechanisms may be restricted to localities. The regional approach is a key element in developing an integrated management program, as it provides a better understanding of ecological interactions and attempts to develop meaningful management strategies.

Five crop situations were selected in which intensive attempts have been made to develop integrated management programs rather than relying primarily on chemical pest suppression. In all of these crop situations there are still many growers using unilateral, shortsighted approaches resulting in target pest resurgence, and outbreaks of secondary pests. The five crops—alfalfa, apple, citrus, cotton, and grape—are leading examples of concerted research and extension efforts in developing and implementing practical pest management programs in the western states. Input from project W-84 has been important in three of the five examples. These programs have been adopted by many growers and are being successfully directed by pest management consultants and agricultural extension personnel.

Operation of Pest Management Programs

Pest management programs are open-ended in nature, allowing for a continuous evaluation, refinement, and reorientation of the management elements. Successful operation of pest management programs depends upon the continuous availability of current information on insect forecasting, sampling and monitoring, and population dynamics, including an understanding of the role played by biotic and abiotic agents. Some decision-making can be accomplished at the farm level if the guidelines are fairly simple and understandable. However, most decision-making involves rather lengthy deliberation and consideration of many elements, including the role played by insecticide resistance and availability of natural mortality factors. Thus, county agents and self-employed pest management consultants must perform a crucial role in the operation of a management program.

One of the critical decisions a grower or pest management operator has to make is to determine whether or not a spray treatment is required at a particular time, and, if so, at what rates and with which chemical. Short-run and long-run economic gains are chief motives for this decision-making process. Some things such as marketing standards, pesticide use regulations, and labor problems are beyond the control of the farmer.

Examples of Pest Management Programs

These examples are selected to illustrate the operation of relatively successful pest management programs. They do not represent the ultimate in successful programs. Some of the methods can be used on other crops, but only after careful study of the biological interactions.

Grape pest management

There are over 600,000 acres of grapes in California, of which well over 70 percent are in the Central Valleys. In 1973, grapes in California were valued at about $609,422,000 (California Farmer, 1974). Three major arthropod species cause severe damage to grapes. These include the grape leafhopper, *Erythroneura elegantula* Osborn, Pacific spider mite, *Tetranychus pacificus* McGregor, and the omnivorous leafroller, *Platynota stultana* Walshingham. The primary disease problem is powdery mildew. Half a dozen other pests occasionally attain a dominant status in some areas but require considerably less amounts of pesticides for their control. The orange tortrix, *Argyrotaenia citrana* (Fernwald), is a problem in the northern San Joaquin Valley grape-growing districts, while the grape leaffolder, *Desmia funeralis* Hübner, can be serious at times throughout the state. Mealybugs, thrips, the grape phylloxera, cutworms, and the Willamette mite, *Eotetranychus willametti* McGregor, are relatively less severe in most vineyards. The pest control program in California varies from complete dependence on natural enemies to complete dependence on chemical pesticides, although the former is rarely practiced in commercial vineyards. An integrated pest management program has been developed by University of California scientists (Flaherty and Huffaker, 1970; AliNiazee and Stafford 1973, 1974; Doutt and Nakata, 1973). Growers commonly use from 1 to 6 treatments per year to control grape pests, with this volume reduced as much as 50 percent under integrated control. Some of the important pest management procedures involving the three major pests are analyzed here.

The grape leafhopper is commonly the most serious pests of grapes in the San Joaquin Valley and causes severe damage annually. During periods of severe outbreaks it has devastated the grape industry. In the Central Valleys of California it completes three generations per year. The overwintering adults appear on vines in April and populations persist until the end of September or early October, at which time they may move out to overwintering shelters. Feeding of adults and nymphs on foliage affects both yield and quality. Leafhoppers produce excessive amounts of excrement on fruit, which adversely effects the appearance of table grapes.

Chemical control has been the major tool of suppression of this pest for the past 25 years. Various insecticides have been used, and replaced by newer compounds usually because of quick development of resistant populations. This is a good example of misuse and overuse of insecticides. Before the commitment to an integrated control philosophy and the adoption of economic injury levels, leafhopper control on vines was a matter of spraying when leafhoppers were seen. In general, growers ignored population density and were treating fields as soon as leafhoppers appeared. Often mixtures of insecticides were employed at frequent intervals. This led to the development of insecticide-resistant leafhoppers and spider mite problems.

The Pacific spider mite is another serious pest of grapes. California's vineyard pest control history suggests a strong correlation between the increased infestation of the spider mites and the excessive use of pesticides for the control of leafhoppers and mites (Flaherty *et al.*, 1973). Flareups of spider mite populations are not clearly understood. The adverse effects of pesticides on the predator, *Typhlodromus occidentalis* Nesbitt, has had a significant influence (Flaherty and Huffaker, 1970). Chemicals like carbaryl (Sevin), dicofol (Kelthane), and azinphosmethyl (Guthion) cause both direct predator mortality and indirect kill by starvation due to lack of prey at critical times. Growers have used numerous pesticides, one after another, because of continuous build-up of tolerance in spider mites to most pesticides. The numbers of spray applications and dosages were repeatedly increased, which aggravated the pest problems rather than solved them.

The omnivorous leafroller is the third serious pest against which chemicals have been only partly successful (AliNiazee and Stafford, 1973). It infests both foliage and fruit and causes extensive damage to the berries. Life history studies indicated that it overwinters as a larvae in trash and grape mummies in the vineyards. Therefore, cultural control should form the foundation, augmented by proper and judicious application of chemical pesticides. Proper timing of insecticide applications can be determined by employing pheromone traps (AliNiazee *et al.*, 1970).

A team effort was initiated during the early 1960's to focus on the development of an effective integrated control program against grape pests in California. At first, the program was supported primarily by the University of California, then later by the grape industry. Efforts by half a dozen scientists during the period 1966-1973 resulted in an integrated control program using a multifactorial approach. This included an understanding of the role played by *T. occidentalis* to suppress spider mites (Flaherty and Huffaker, 1970), cultural controls and stomach poisons to suppress omnivorous leafrollers (AliNiazee et al. 1970; AliNiazee and Stafford, 1973), development of economic injury levels for the grape leafhopper (Jensen *et al.*, 1973), utilization of a parasite, *Anagrus epos* Girault, for leafhopper control (Doutt and Nakata, 1973), and *Bacillus thuringiensis* Berliner to control grape leaffolders (AliNiazee and Jensen, 1973).

Basically, the program involves delaying the use of insecticides to control grape leafhoppers until the population reaches an economic threshold level of 20 nymphs per leaf in the first brood, or 15 nymphs per leaf in the second brood on Thompson seedless grapes for raisins or wine (Jensen *et al.*, 1973). It preserves *A. epos* to control the leafhopper by reducing the use of insecticides during April and May when the parasites occur in large numbers. Delaying early season use of insecticides also allows *T. occidentalis* populations to increase. The program includes cultural controls, the early season use of stomach poisons like cryolite to suppress *P. stultana*, and pheromone traps to determine proper timing of late-season application of organic insecticides. It suggests the use of *B. thuringiensis* for grape leaffolder control. If properly followed, 2 to 3 spray treatments can be eliminated each year, with a $40 to $60 per acre savings to the growers.

It is estimated that 10 to 20 percent of the San Joaquin Valley vineyards are partly or completely under this program. Increased availability of qualified pest management consultants, plus grower awareness, will improve this figure considerably. If more efficient and simple population monitoring techniques become available the acceptance of the program will improve.

Citrus pest management

Pest management on citrus in California is generally considered to be the classical example of biological control. There has been a long history of use of natural enemies against citrus pests in California. This background served as the foundation for pest management development. About 54 percent of the California citrus industry is in the southern portion of the state, divided between coastal and interior areas. Most of the remaining, much of which are newer plantings, occurs in the San Joaquin Valley (Deal, 1972).

In the coastal areas of California the two primary pests of citrus are the citrus red mite, *Panonychus citri* (McGregor), and the California red scale, *Aonidiella aurantii* Maskell. Minor pests include the citrus rust mite, *Phyllocoptruta oleivora* (Ashmead), citrus bud mite, *Aceria sheldoni* (Ewing), European brown snail, and miscellaneous aphids, mealybugs, and orangeworms. In the intermediate and interior areas, primary pests are the citrus red mite, California red scale, and citrus thrips, *Scirtothrips citri* (Moulton). Minor pests include aphids, orangeworms, yellow scale, *Aonidiella citrina* (Coquillett), and citricola scale, *Coccus pseudomagnoliarum* (Kuwana), (Deal, 1972).

Pests which can be kept under control by natural enemies are the brown soft scale, *Coccus hesperidum* L., black scale, *Saissetia oleae* (Bernard), and cottony cushion scale, *Icerya purchasi* Maskell (Deal, 1972).

Control of the latter pest in California in 1888-89 by the vedalia ladybeetle, *Rodolia cardinalis* (Mulsant), and a parasitic fly, *Cryptochaetum iceryae* (Williston), was the first and probably the most spectacular success in biological control (DeBach, 1974).

Pest control programs for citrus in California vary from complete dependence on natural enemies to almost total chemical control. Recently, however, more and more growers have been using an integrated approach to pest control, now more popularly called pest management. The basic concept is not new, having been practiced by the Fillmore Citrus Protective District in Ventura County for more than 50 years. This district, comprising about 8,800 acres, is a cooperative pest control association under the supervision of a trained entomologist. It maintains its own insectaries to produce natural enemies as needed, and uses only a few selective insecticides (DeBach, 1974).

Altogether, ten insect pests of citrus are being controlled satisfactorily by their natural enemies in the Fillmore district. These are the California red scale, black scale, yellow scale, brown soft scale, cottony cushion scale, dictyospermum scale, and four species of mealybugs. Pests which are partially controlled by native natural enemies are the citrus red mite, aphids, and orangeworms (DeBach, 1974). In addition to emphasizing biological control, the district discourages the use of broad-spectrum insecticides. Selective pesticides are sometimes applied as spot treatments against scales, mites or thrips. The Fillmore district has a highly successful integrated control program which provides low pest-control costs and high quality and quantity of fruit.

Apple pest management

The leadership in developing a pest management system for use in apple orchards has come from the state of Washington. However, with certain modifications, the concept has been accepted widely in most western states and British Columbia. The discussion here centers on how it has been developed for use in Oregon and Washington.

The codling moth, *Laspeyresia pomonella* (Linn.) is the key pest in the western United States. It attacks the fruit causing stings, entrance holes, and the well known "wormy apple." Growers use from two to five applications of organophosphorous or carbamate insecticides each season to control the codling moth. These chemicals traditionally are used as preventive or insurance treatments and begin soon after emergence of the first codling moth adults. Although most codling moth populations are apparently still susceptible to currently available insecticides, the possibility of this insect's becoming resistant to insecticides cannot be overlooked (Hoyt and Burts, 1974). Because chemical control of codling moths is dependable, most growers are not interested in alternative methods. Although several nonchemical practices can reduce codling moth populations no extensive use is made of them. Growers are not receptive to cultural practices which mean additional work. Backyard trees and abandoned orchards are important sources of codling moth infestations in many orchards. The removal of abandoned trees in the vicinity of commercial orchards can reduce sources of codling moths.

A complete dependence on codling moth control by chemicals is not without problems. Hoyt (1969) and Hoyt and Caltagirone (1971) discussed this in detail. Most orchards with excessive use of pesticides have experienced severe spider mite outbreaks. In addition natural enemies of other pests were eliminated, resulting in outbreaks of San Jose scale, apple aphid, leafrollers, and other pests normally kept under control by natural enemies.

The McDaniel spider mite, *Tetranychus mcdanieli* McGregory, the European red mite, *Panonychus ulmi* (Koch), and the apple rust mite, *Aculus schlectendali* (Nalepa), are the most common foliage feeding mite species in Washington (Hoyt, 1969). In the Hood River Valley of Oregon, the twospotted spider mite, *Tetranychus urticae* (Koch), McDaniel mite and European red mite are most common (Zwick, 1972). McDaniel mites develop resistant populations rapidly (Hoyt, 1969). As one acaricide failed, others soon replaced it, leading to a long history of diverse materials. This did not solve the problem, but rather, intensified it. A more stable alternative mite control technique became essential.

In the Willamette Valley of Oregon the major mite species are the yellow spider mite, *Eotetranychus carpini borealis* (Ewing), and the European red mite. Mixed populations of these two species and the rust mite and McDaniel mite occur in many orchards (AliNiazee, 1978). Generally, the use of acaricides has been much less intense and consequently many of the older acaricides are often still effective. However, in orchards where a calendar spray program has been followed, the situation is not much different than what is encountered in Washington.

The mite control problems in western Colorado, Idaho, western Montana, and Utah followed similar patterns to those of Washington with routine acaricide applications.

Hoyt (1969) recognized the important role played by the phytoseiid mite, *T. occidentalis*, in apple orchards in Washington. He noticed that in one of the study orchards this mite survived a regular spray program and effectively suppressed spider mites. This led to the discovery of Guthion-tolerant strains of *T. occidentalis*. Later, effective rates of Guthion and some other organophosphate insecticides were determined, which obtained good control of codling moths while sparing most predatory mites. Further studies indicated that the apple rust mite plays a key role in the survival and effectiveness of *T. occidentalis* by providing an alternate food source, particularly during the periods when the McDaniel mites are in low numbers. These discoveries led to the development of an integrated control program on apples in Washington and in other western states.

This program, discussed in detail by Hoyt and Caltagirone (1971), involves a pre-bloom spray to control European red mite eggs, San Jose scales, early emerging lepidopterous larvae and aphid eggs. A proper timing of early season codling moth sprays determined by pheromone traps is essential. These sprays should include materials to control San Jose scale crawlers, and apple aphids. Later sprays (after May-June) are recommended only when codling moths or other pests warrant treatment. Late-season sprays are more detrimental than early sprays to *T. occidentalis* populations and the predator:prey balance. If early season spray guidelines are properly followed, and *T. occidentalis* populations are sufficiently high and synchronized with the prey, then mite control is not needed during July and August.

This integrated control program has been widely accepted by growers in Washington because of its effectivess in reducing mite problems with a dramatic reduction in the cost of pest control. It has been estimated that an $85 per acre cost for pesticides alone during the peak pesticide use period was reduced to about $25 per acre when the integrated control program was followed (Hoyt and Caltagirone, 1971). The number of growers in the program has fluctuated annually, with many being enthusiastic and following the recommendations of integrated control advisors.

In the Hood River Valley of Oregon, attempts to develop a similar program utilizing *T. occidentalis* were less successful. *T. occidentalis* was incapable of maintaining spider mites below economic injury levels, probably because fewer rust mites were present. By employing selective acaricides, however, increased befits from predatory mites were obtained (Zwick, 1972).

In the Willamette Valley of Oregon, another potentially useful phytoseiid mite, *Typhlodromus arboreus* (Chant) was noted in the fall of 1972 (AliNiazee, 1978). Later studies indicated that under favorable predator: prey balance conditions this mite is capable of holding the spider mites below economic injury levels. In one apple orchard where this mite was eliminated by a carbamate insecticide, the spider mite population reached an alarming density and caused virtually defoliation of the trees (AliNiazee, 1978). Attempts are being made to develop an integrated control program using *T. arboreus* and a low rate of Guthion with *Bacillus thuringiensis*. This program involves an application of a dormant spray and two summer sprays to control codling moths (AliNiazee, unpublished data); the codling moth sprays must be properly timed with the aid of pheromone traps. No other insecticides or miticides are recommended during the summer months, thus allowing predatory mite populations to increase enough to suppress spider mite populations.

Alfalfa pest management

Some of the earliest interest in pest management was in supervised control of the alfalfa caterpillar and integrated control of the spotted alfalfa aphid in California. Since that time, interest has spread to other pest species, and pest management in various stages of development is being used in many alfalfa fields throughout the United States. The system used in California can serve as a model.

The principal insect pests of alfalfa in California are the alfalfa weevil, *Hypera postica* (Gyllenhal), the Egyptian alfalfa weevil, *Hypera brunneipennis* (Boheman), the alfalfa caterpillar, *Colias eurytheme* Boisduval, the spotted alfalfa aphid, *Therioaphis trifolii* (Monell), pea aphid, *Acrythosiphon pisum* (Harris), western yellow-striped armyworm, *Prodenia praefica* Grote, beet army-worm, *Spodoptera exigua* (Hübner), and webworms, *Loxostege* spp. (Stern, 1966). Recently the blue alfalfa aphid, *Acyrthosiphon kondoi* Shinji, has been added to the list. Economic threshold levels (based on direct counts) have been established for all of these pests except the blue alfalfa aphid, western yellow-striped armyworm, and webworms.

The alfalfa weevil occurs in the cooler northern areas of California, and the Egyptian alfalfa weevil in the warmer areas of the state. Both are introduced (exotic) pests which primarily injure alfalfa in winter or cool spring. An introduced species, *Bathyplectes curculionis* (Thomson), is an excellent parasite of the alfalfa weevil, having been released against the pest in California during 1933 and 1934 (Michelbacher, 1943). The parasite readily became established and is most effective against the alfalfa weevil in the coastal regions. Though several imported parasites have been released in the San Joaquin Valley of California during recent years against the Egyptian alfalfa weevil, only a pteromalid, *Dibrachoides druso* (Walker), has become established. The species was recovered in San Diego County in 1967 about 5 years after its release (Gonzalez *et al.*, 1969), but was not considered to be an effective parasite of the pest in that area.

The alfalfa caterpillar is a native pest in California which is normally limited in abundance by natural enemies, especially by a braconid wasp, *Apanteles medicaginis* Muesebeck, an egg parasite, *Trichogramma semifumatum* (Perkins), and a nuclear polyhedrosis virus which attacks the larvae (Stern, 1966). *A. medicaginis* is the key mortality factor in regulating the alfalfa caterpillar in the Central Valleys. Both *A. medicaginis* and *T. semifumatum* are major biological control agents of the pest in southern California (Stern and Bowen, 1963). The polyhedrosis virus is especially effective in reducing the larval populations during late summer (Michelbacher and Smith, 1943).

Michelbacher and Smith (1943) developed a simple technique of evaluating parasitism of the alfalfa caterpillar by *A. medicaginis* which involved field dissection of small larvae, parasitized ones being easily detected. If more than 10 larvae per sweep were unparasitized, a selective insecticide such as *Bacillus thuringiensis* Berliner var. *thuringiensis* was recommended. The technique was first used by California alfalfa growers in 1946 under the supervision of a qualified entomologist. The supervised control system was developed further and became a prelude to the development of a sophisticated integrated control approach to pest problems (Hagen *et al.*, 1971).

The spotted alfalfa aphid invaded California in 1954 and soon became a major pest, causing severe damage to alfalfa (Hagen and van den Bosch, 1968). Three parasites, *Praon palitans* Muesebeck, *Trioxys utilis* Muesebeck, and *Aphelinus semiflavus* Howard, were imported from the Middle East and released against the pest in California during 1955 and 1956. By 1959, the parasites were well established in most alfalfa-growing districts. In addition to the three imported parasites, native predators (particularly coccinellids), an entomogenous fungus, and aphid-resistant varieties of alfalfa help keep the spotted alfalfa aphid below levels of economic injury (Stern, 1966). A selective insecticide (demeton) was used while biological control and resistant varieties of alfalfa were being developed.

In central and northern California, the pea aphid is kept under biological control (especially in coastal fields of alfalfa) by predators, entomogenous fungi, and parasites such as *Aphidius pulcher* Baker and *A.*

smithi Sharma and Rao (Hagen *et al.*, 1971). The latter species was imported from India in 1958 and released in California where it quickly became established (Hagen and van den Bosch, 1968). By 1960, *A. smithi* had spread over much of the alfalfa-growing areas and had become an important controlling agent of the pea aphid in coastal valleys. Its effectiveness is greatly increased in strip-harvested fields of alfalfa.

The western yellow-striped armyworm, beet armyworm, and webworms attack alfalfa during the summer months (Stern, 1966). These native pests have a number of hymenopterous and dipterous parasites which help keep them under control, especially in strip-harvested fields. These natural enemies are conserved by the use of proper timings and dosages of selective insecticides if treatments are required for any alfalfa pest.

Cotton pest management

As with alfalfa and apples, the original development of pest management came from one area. The system as developed in California is used as the example, but pest management of cotton pests is being employed in some form in most cotton-growing states.

The major insect pests of cotton around which a pest management system is built in the San Joaquin Valley are lygus bugs (mostly *Lygus hesperus* Knight) and the bollworm, *Heliothis zea* (Boddie). Pest problems which often follow insecticide treatments include spider mites (*Tetranychus* spp.), the beet armyworm, *Spodoptera exigua* (Hübner), cabbage looper, *Trichoplusia ni* (Hübner), salt-marsh caterpillar, *Estigmene acrea* (Drury), stink bugs, and aphids (Reynolds 1969).

Prior to 1965 the major insect pests in the Imperial and neighboring low desert valleys were lygus bugs and the bollworm. Occasional (minor) pests were the cotton leaf perforator, *Bucculatrix thurberiella* Busck, cabbage looper, beet armyworm, salt-marsh caterpillar, omnivorous leaf roller, and spider mites. In the early 1960's integrated control was successfully practiced by supervising entomologists who evaluated populations of insect pests in each field and recommending applications of selective pesticides only when biotic mortality factors (parasites, predators, and pathogens) appeared inadequate (Reynolds, 1969). In 1965 the pink bollworm, *Pectinophora gossypiella* (Saunders), invaded the area and caused severe crop losses. To combat this new key pest of cotton, growers resorted to frequent applications of broad-spectrum organophosphate insecticides. As a result, pest control costs increased sharply and populations of natural enemies were almost totally eliminated. Several normally minor pests, such as the cotton leaf perforator, soon reached destructive levels (Reynolds, 1969). Now, a decade later, cotton growers have become aware that their survival is dependent on regaining a pest management approach to control.

Much has been learned about pest sampling and prediction methods, economic thresholds and injury levels, the role of naturally occurring biotic mortality ages (natural enemies) and their manipulation for suppression of pest populations, impact of pesticides on natural enemies, manipulation of pest populations through cultural and agronomic practices, and the development of ecologically selective pesticide controls. Currently, all of these techniques are being used in developing an integrated control program for cotton pests in the San Joaquin Valley (van den Bosch *et al.*, 1971).

In spite of considerable research efforts over a long period of time, reliable economic thresholds have not been established for the two major pests, *L. hesperus* and the bollworm. Nevertheless, much progress has been made, especially on the former species (van den Bosch *et al.*, 1971). It has been determined that extensive early and midseason insecticide treatments for control of lygus bugs is responsible for most subsequent outbreaks of lepidopterous pests and spider mites. Therefore, a more flexible treatment threshold has been developed based on pest density, especially during the period from pre-bloom to 2 weeks after bloom initiation (June through mid-July) (van den Bosch et al., 1971).

Cultural methods of reducing lygus bug numbers in the San Joaquin Valley cotton fields include strip-cutting of nearby alfalfa fields, interplanting strips of alfalfa in cotton fields, and properly-timed applications of insecticides for control of lygus bugs in maturing safflower fields (Stern *et al.,* 1964; Stern *et al.,* 1969).

The bollworm is rarely abundant on cotton in the San Joaquin Valley before early August. Moreover, in untreated cotton, natural enemies take a heavy toll of bollworm larvae. This includes predators such as *Geocoris* spp., *Nabis* spp., and *Chrysopa* spp. (van den Bosch *et al.*, 1971). Most insecticide treatments are destructive to these natural enemies. Thus, much of the control by the chemicals only replaces that which would have been accomplished by predators and parasites. To help reduce the number and frequency of chemical applications against the bollworm, the population level for treatment was increased from four bollworm larvae per 100 plants to a provisional level of 15 (van den Bosch, *et al.*, 1971).

As with all successful pest management systems, the program on cotton in the San Joaquin Valley will require close monitoring of fields by technically trained individuals. By 1971 only about 10 percent of the cotton acreage was being supervised, but acreage was growing as qualified persons became available to supervise the program (van den Bosch *et al* ., 1971).

Literature Cited

ALINIAZEE, M. T.

1978. Role of a predatory mite, *Typhlodromus arboreus* in natural control of spider mites on apples in western Oregon. Proc. IV International Congress Acarology, Saalfelden, Austria, 1974. (In press.)

ALINIAZEE, M. T., C. D. LYNN, E. M. STAFFORD and D. A. LUVISI.

1970. Integrated control of grape pests: effectiveness of cryolite and standard lead arsenate against the omnivorous leafroller. Calif. Agric. 24(12): 8-10.

ALINIAZEE, M. T. and C. L. JENSEN.

1973. Microbial control of the grape leaffolder with different formulations of *Bacillus thuringiensis.* J. Econ. Entomol. 66: 157-58.

ALINIAZEE, M. T. and E. M. STAFFORD.

1971. Evidence of a sex pheromone in the omnivorous leafroller: laboratory and field testing of male attraction. Ann. Entomol. Soc. Amer. 64: 1330-35.

1973. Management of grape pests in central California vineyards. I. Cultural and chemical control of *Platynota stultana* on grapes. J. Econ. Entomol. 66: 154-57.

1974. Management of grape pests in central California vineyards. II. Toxicity of some commonly used chemicals to *Tetranychus pacificus* and its predator *M. occidentalis.* J. Econ. Entomol. 67: 543-48.

CALIFORNIA FARMER.

1974. 1973 State agricultural production valued at $7.6 billion. Calif. Farmer 240: 37-38.

DEAL, A. S.

1972. Present status of pest management systems in citrus, pp. 154-58. *In* Implementing Practical Pest Management Strategies. Proc. Nat. Exten. Insect-Pest Management Workshop. Purdue University, Lafayette, Indiana. 206 pp.

DEBACH, P.

1974. Biological Control by Natural Enemies. Cambridge University Press. London-New York. 323 pp.

DOUTT, R. C. and J. NAKATA.

1973. The *Rubus* and its egg parasitoid: an endemic biotic system useful in grape pest management. Environ. Entomol. 2: 381-86.

FLAHERTY, D. L. and C. B. HUFFAKER.

1970. Biological control of Pacific mites and Willamette mites in San Joaquin Valley vineyards. I. Role of *Metaseiulus occidentalis.* II. Influence of dispersion patterns of *M. occidentalis.* Hilgardia 40: 267-330.

FLAHERTY, D. L., C. D. LYNN and F. L. JENSEN.

1973. Pest management of spider mites of grapes in the southern San Joaquin Valley. U.C. San Joaquin Valley Agricultural Research and Extension Center. Publi. 25 pp.

GONZALEZ, D., R. VAN DEN BOSCH, and L. H. DAWSON.

1969. Establishment of *Dibrachoides druso* on the Egyptian alfalfa weevil in southern California. J. Econ. Entomol. 62: 1320-22.

HAGEN, K. S. and R. VAN DEN BOSCH.

1968. Impact of pathogens, parasites, and predators on aphids. Ann. Rev. Entomol. 13: 325-84.

HAGEN, K. S., R. VAN DEN BOSCH, and D. L. DAHLSTEN.

1971. The importance of naturally-occurring biological control in the western United States, p. 253-293. *In* C. B. Huffaker (ed.), Biological Control. Plenum Press, New York and London. 511 pp.

HOYT, S. C.
1969. Integrated chemical control of insects and biological control of mites on apples in Washington. J. Econ. Entomol. 62: 74-78.
HOYT, S. C. and E. C. BURTS.
1974.. Integrated control of fruit pests. Ann. Rev. Entomol. 19: 231- 52.
HOYT, S. C. and L. E. CALTAGIRONE.
1971. The developing program of integrated control of pests of apples in Washington and peaches in California. pp. 395-421. *In:* Huffaker, C.B. (ed.), Biological Control, Plenum Press, New York. 511 pp.
JENSEN, F. L. FLAHERTY, C. D. LYNN, and R. L. DOUTT.
1973. Grape pest management in the southern San Joaquin Valley: grape leafhoppers. U.C. San Joaquin Valley Agricultural Research and Extension Center. Publi. 26 pp.
MICHELBACHER, A. E.
1943. The present status of the alfalfa weevil in California. Calif. Agr. Expt. Sta. Bul. 677. 23 pp. (out of print).
MICHELBACHER, A. E. and R. F. SMITH,
1943. Some natural factors limiting the abundance of the alfalfa butterfly. Hilgardia 15(4): 369-97.
MUELLER, A. J. and V. M. STERN.
1974. Timing of pesticide treatments on safflower to prevent *Lygus* from dispersing to cotton. J. Econ. Entomol. 67: 77-80.
REYNOLDS, H. T.
1969. Insect and mite control problems and research in California. Proc. Cotton Symposium. University of California, Berkeley. 146 pp.
STERN, V. M.
1966. Significance of the economic threshold in integrated pest control. Proc. FAO Sympos. Integrated Pest Control, Rome, Italy, 2: 41-56.
STERN, V. M. and W. R. BOWEN.
1963. Ecological Studies of *Trichogramma semifutum* with notes on *Apanteles medicaginis* and their suppression of *Colias eurytheme* in southern California. Ann. Entomol. Soc. Amer. 56: 358-72.
STERN, V. M., A. MUELLER, V. SEVACHERIAN, and M. WAY.
1969. *Lygus* bug control in cotton through alfalfa interplating. Calif. Agr. 23(2): 8-10.
STERN, V. M., R. VAN DEN BOSCH, and T. F. LEIGH.
1964. Strip cuttin alfalfa for *Lygus* bug control. Calif. Agr. 18(4): 4-6.
VAN DEN BOSCH, R., T. F. LEIGH, L. A. FALCON, V. M. STERN, D. GONZALEZ, and K. S. HAGEN.
1971. The developing program of integrated control of cotton pests in California, p. 377-93. *In* C.B. Huffaker (ed.), Biological Control. Plenum Press, New York and London. 511 pp.
ZWICK, R. W.
1972. Studies on the integrated control of spider mites on apples in Oregon's Hood River Valley. Environ. Entomol. 1: 169-76.

Chapter 11.

Developing A Pest Management Program

T. F. Watson, G. W. Bishop and L. Moore

Introduction

Before a pest management program can be developed and initiated a great deal of basic information must be accumulated. This includes information about the agroecosystem (such as the crops grown), agronomic practices employed, soil type, irrigation water, and any other factors related to production of crops. Detailed information must be available on the major pests and beneficial insects found in the agroecosystem in order to understand the seasonal occurrence and magnitude of all species of concern. Integration of all information on the agroecosystem with information on the biology and ecology of pests and beneficial insects will provide significant insight on natural control in any area. The level of natural control provides the base on which all management practices are built.

Two additional elements basic to the development of a pest management program are sampling and economic injury levels. A good sampling system is extremely important as it provides information on insect numbers in each field and must be developed to serve as a base for utilizing knowledge of natural control, economic injury levels, and biology and ecology of the major insects involved. Once the sampling program is established, these basic elements can be integrated to serve as the foundation upon which practical components can be added to the total pest management program (Watson *et al.*, 1976)

The following discussion pertains primarily to a large agroecosystem involving any or all of the major agronomic or horticultural crops produced in the area. It is also intended to encompass the needs of a number of growers, forming a more or less compatible unit in terms of similarity of crops grown and mutual pest problems encountered. It does not preclude, however, the development of equally effective pest management programs on individual farms or on more specialized crops. The basic requisites would still be the same and the discussion which follows would apply equally as well, with appropriate modifications to meet special needs. For example, an individual grower would still need the services of a pest management specialist to provide the basic information on which sound pest control and management decisions could be made. The organization and responsibilities remain the same, but are less complex.

Organizational Structure

Organizational units involved

The organizational structure necessary to establish and operate a pest management program involves a growers' committee or organization, advisory personnel—usually university extension service personnel—supervisors, and field samplers or "scouts." A knowledgeable, progressive grower organization is the key to establishing a pest management program. Without enthusiastic grower support and participation a pest management program has little chance of succeeding.

An institutional unit such as an agricultural extension service plays a vital role in the development of a pest management program. However, the function of the institutional unit should be that of a steering or advising committee. Extension service personnel can provide the technical information needed by the grower committee to properly organize the entire program. It is important, however, that the grower committee retain leadership of the program in order to avoid the connotation of its being a government- or university-sponsored program. It also leaves no doubt as to the responsibility by the growers for the conduct of the program.

In addition to extension entomologists, advisory personnel include research entomologists, county agents or farm advisors, and other entomology professionals available to the organization. Some contact with the grower committee by all the above-mentioned advisory personnel provides the grower committee with more complete information relative to their problems and thus enhances the over-all program.

Supervisors are trained entomologists or persons experienced in conducting pest management programs. This group is an essential link between the growers and the field samplers or scouts.

Field samplers or scouts are primarily college students, although they may include school teachers, high school students and persons employed specifically for this purpose on a year-round basis. Usually they are hired locally and for the period when crop scouting is required. These individuals are usually less knowledgeable in areas such as population dynamics, ecology, and pest management strategies, but are trained specifically for sampling certain pest and beneficial species in a prescribed manner and for providing the continuity on a week-to-week basis of population levels in every field. The utilization of these population assessment data is left to the supervisors and the growers.

Responsibilities of various units

A committee or board, composed of seven or more growers, should be selected to head the grower organization. The primary responsibility of this committee is to represent the entire organization on all matters of policy and to see that the program runs in an efficient and effective manner. The committee is also responsible for: meeting with the advisory personnel and deciding on courses of action, making financial arrangements for conducting the program, hiring supervisors and assigning them to program contracts, obtaining the necessary insurance on personnel and crop loss, and meeting regularly to discuss and evaluate the program. In arranging financing it may be advantageous for the growers to work with existing organizations such as grower cooperatives, gin groups, or irrigation districts.

Advisory personnel are responsible for assisting the grower committee or board in organizing the program, training the field scouts in insect identification and techniques of field sampling, assisting with weekly meetings of scouts, supervisors and growers to discuss problems and situations, and providing the growers with information on insects and their control. Continuous contact of advisory personnel with the supervisors is important and provides many benefits to both parties. It is advantageous to the advisors in that they are better informed on the insect situation in a localized area. In turn, they can bring broader knowledge to the supervisors concerning pest situations from outside the program area. Additionally, they can apprise the supervisors of new research findings that will assist them in their decisions, and give first-hand assistance to the supervisors on new or difficult situations.

The supervisor's first responsibility is to line up a good, dependable field force that can provide him with the necessary information for decision-making. After hiring the scouts and making the necessary arrangements for their training , he must assign scouts to an area and fields within that area for which they are responsible. He is also responsible for providing the scouts with field-record forms and other supplies which they need, discussing field-record information with growers, rechecking fields having borderline infestations, and assisting growers in identifying and applying on a timely basis the most practical and desirable pest management tactics available.

A few special preparations are essential in initiating a good sampling program and the supervisor is usually responsible for these. It is necessary that he obtain aerial photo-maps that clearly show the fields to be sampled. He must then number the fields for identification, using a four-digit numbering system (this is needed to provide a reserve of potential numbers to permit indefinite assignment of field numbers). He should assign the fields to the scouts while permitting them to study both appropriate aerial and road maps to locate the fields prior to or during the first week of sampling.

The field scout is responsible for properly sampling and recording information weekly or as often as required on insect conditions in assigned fields, providing completed record sheets to the grower or supervisor as arranged, and attending training sessions, including weekly meetings. His total responsibility is to know what is going on in the fields to which he is assigned.

Funding

In order to acquaint the grower with pest management and to provide the initial thrust, pilot programs were funded by the federal government, first in Arizona on cotton and in North Carolina on tobacco. They were then expanded to include a number of different crops and cotton in 14 states. Prior to this time private consultants, particularly in California, had ongoing supervised control programs which were essentially pest management programs.

When the federal government became involved, grants were made available to various state extension services across the country to provide financing for leadership of these programs and for other costs, including costs of sampling the fields. The ultimate goal was to increase the annual cost to the grower until the entire cost of such programs was being borne by the grower.

In private grower organizations, the grower committee sets the policy as to cost per acre, determines when and how the funds are to be collected, and designates the person responsible for collections and payments on expenses. The organization usually has an account, and the officer designated to oversee this account (perhaps a person such as a gin manager) pays supervisors, scouts, and bills as they become due. Salaries and methods of payment are set prior to the hiring of the personnel each season.

Data Processing

Immediate use

The main purpose of sampling and population assessment is to provide data needed for making control decisions. In many crops the status of a pest infestation can change quite rapidly, particularly with migratory species that may leave one crop and move to another. Therefore, it is essential that sample data be available to the growers within hours after they are collected if pest levels are such that an immediate threat to the crop exists. Many of the larger growers are difficult to locate on a day-to-day basis, so a pre-arranged drop-point is established where the scout can leave the field-record sheet soon after the samples are taken. The grower knows the day of the week and time of day that he can expect to find his records at the drop-point and can pick them up for assessment of his situation. He then makes decisions on control or he can call the supervisor for assistance in interpreting the data.

A copy of each field-record sheet is delivered daily to the supervisor so that he can review the pest and beneficial insect situation for fields sampled on that day. If records indicate certain fields with pest levels exceeding the economic thresholds, the supervisor contacts the grower to alert him of the situation. If they indicate levels approaching the economic thresholds, he personally re-checks the field the following day in order to verify the sample data and to better advise the grower on the course of action to take, if requested.

The sample data are also utilized by the scouts to determine the total picture of the pest situation in their area of responsibility. The data may indicate that a particular pest of one crop has reached damaging levels only in areas associated with other crops. Unusual trends of pest populations in one crop may be explained by the grower practices in an adjacent crop; e.g., the cutting of an alfalfa field will usually result in the movement of high numbers of lygus bugs into adjacent cotton.

One additional immediate use of the field records is for discussion and questions at the weekly meetings of scouts, supervisors, advisors, and possibly growers. This provides each scout with a broader perspective of what is happening in the entire area other than what each has been observing on a day-to-day basis in his particular fields. It also provides a chance for the supervisors or advisors to clarify unusual occurrences, and to alert the scouts to potential developments.

Historical use

Some pest management programs have computerized data from field-record sheets to provide a permanent data-bank that is available for use at any time. This approach has provided information printouts on weekly population trends, geographical influence on pest populations, crop association with various pests, and insecticide applications and pest and beneficial insect trends.

Another important use of the permanent data bank is to provide information for use in research. Many interesting observations are made, relative to these historical data, which need verification through experimentation.

Growers' Planning and Decision Making

The cropping system in relation to pest management

The cropping system is extremely important in a pest management program. Generally, a diversified type of agriculture is desirable because it provides more checks and balances of beneficial insects in the system. It also reduces the possibility of catastrophic outbreaks of pests which are more likely to occur in large monoculture plantings. Even with large operations where extensive monoculture exists, there are usually several crops involved. Generally, one or two are the primary cash crops and the others are utilized for supplemental income, to compliment the rotation system, or as "soil-builders."

In planning their cropping system, growers may utilize information on the biology and ecology of both pests and beneficial insects to enhance pest management. For example, several important species of cutworms prefer to oviposit in a sod crop. If a row crop is planted after a sod crop, chances of severe damage to the young seedlings are great. The practice of destroying weeds and other plants along ditch banks, field borders, etc., eliminates the nectar source of parasites and thus reduces effectiveness of biological control of pests in the planted crop. As an example, grain sorghum tends to build large predator populations and if planted adjacent to cotton provides a reservoir of predators that may move into the cotton.

A small manipulation of routine production practices of the cash crop itself will sometimes improve management capabilities of certain pests. For example termination of the cotton crop at an earlier-than-normal date in the fall has been shown to be an effective practice in the management of the pink bollworm.

Decision making

Rather than handling each crop as a separate entity relative to pest control, growers can enhance their pest management program if they consider the total agroecosystem as a unit. Decisions concerning pest control on one crop may significantly affect pest problems on other crops. On the western side of the San Joaquin Valley of California where safflower and cotton are the two main crops, the method of coping with the lygus bug problem is extremely important and interrelated between the two crops. Lygus cause no economic damage to safflower, but large populations build up on this crop and then move to cotton as the safflower matures. This happens at the time when cotton is fruiting heavily and severe damage occurs unless insecticide treatments are made. These applications, in turn, trigger the outbreak of other pests such as cabbage looper, bollworm, and spider mites, which require additional insecticide applications. Mueller and Stern (1974) have shown that one properly timed insecticide application for lygus reduced the need for control of lygus on cotton and thus minimized the chance for additional pest outbreaks. In addition to the immediate economic benefits from handling a pest problem in this manner, certain long-term effects could be realized. Two of the more important ones are less insecticide contamination of the environment and a delay in the development of resistance of all pests because of the need for fewer insecticide applications.

Similar benefits can be obtained by growers who have both alfalfa and cotton in their farming operation. Again lygus bugs are the key insects. Stern *et al.*, (1964) showed that lygus prefer alfalfa to cotton and that, where succulent alfalfa is maintained in the area, the lygus remain in the alfalfa. They developed a strip-cutting system in which alternate strips of alfalfa were cut at 2-week intervals, causing the lygus to move back and forth within the alfalfa field. When this was practiced, no insecticide control of lygus on cotton was needed. Using this method, Rakickas and Watson (1974) also showed that certain important predators can be maintained in the area throughout the summer, whereas populations decline fairly sharply by early summer where the solid-cutting method is used.

Where growers utilize the pest management approach, long-term benefits will ensue. By utilizing cultural practices or crop manipulation to suppress pest species or minimize their damage, less insecticidal control is required. This in turn improves conditions for the maintenance of predators and parasites in the area, and thus improves biological control.

Growers in pest management programs are generally in close contact with the research and extension entomologists by virtue of their contact with the supervisors and field scouts. New information on economic thresholds, more effective or less disruptive control measures, insect resistance, etc., is readily available and can be utilized almost immediately by the grower. This permits rapid changes in their over-all strategy in managing their insect pests. Ideally, it is a dynamic system with new information being incorporated almost continuously.

Essentials for Maintaining Commercially-Used Pest Management Programs

Grower requirements

The first requirements in establishing and maintaining a commercially accepted pest management program is to educate growers on the philosophy, techniques, and economic advantages of pest management. From the grower's standpoint the program must be effective and economically sound. If the program can meet these conditions it will become an integral part of his production practices.

The effectiveness and economic advantages of pest management systems can best be documented by maintaining good records of costs, pest and natural enemy populations, weather, irrigation and other important factors which influence crop production. The causes of reduced yields or quality can more readily be identified and steps taken to correct the problem when these records exist.

The second requirement is the availability of pest management specialists who are thoroughly familiar with the technical information necessary to manage the pests on all crops in the system. This includes extensive knowledge of pests, beneficial insects, and the growth characteristics and production practices on each crop. Displaying this knowledge is not only essential to managing the pests but to gaining the confidence of growers. To be most effective, the specialist should have no financial interest in the program other than obtaining the optimum in cost: benefit ratios for the grower. This insures the free choice to utilize all available management tactics.

The economical aspect of the program can be realized in a number of ways. It may simply be in the form of saving on insecticide costs. If one needless application of insecticide can be eliminated, the additional cost of the consulting service is usually more than paid for. Other ways of reducing costs are: (1) more precise timing of insecticide applications: (2) better choice of insecticides for the problem(s) at hand; (3) alternative ways of coping with certain pest problems which may require no additional outlay of capital: and (4) higher quality products as a result of more careful management of the pests.

The program must be effective, and this can be demonstrated by the careful and extensive keeping of records. Often, any reduction in yield or quality of an agricultural crop is attributed to insect damage. If careful records have been maintained on population assessments, climatic factors, irrigation, tillage practices, and any other factors which may have affected the crop, the probable influencing factor may be pinpointed.

The effect of many pest management practices on certain pests may be readily observed after application of the practice. It will probably be wise and necessary for the pest management specialist to call these events to the attention of the grower as they happen. For example, the effectiveness of biological control on a lepidopterous pest can be readily demonstrated by showing egg and larval counts to the grower, illustrating that the potential problem is not developing as a result of predation and parasitism.

Closely allied to economy and effectiveness is reliability of the program. If it has been adequately demonstrated to the grower that the program is both economical and effective, the reliability part will generally take care of itself. The more the grower can be acquainted with the program and apprised of the ongoing situations, the more reliable it becomes.

The pest management specialist

The pest management specialist must be thoroughly familiar with technical information necessary to manage pests on all crops in the system. This means not only extensive knowledge of pests and beneficial insects found in the system but also the growth characteristics and production practices pertaining to the crops.

The specialist should be able to work with growers of differing personalities, and should be the type that instills confidence in growers. The demonstration of his technical competence on a continuing basis will soon erase any lack of confidence which the growers may have had at the outset.

Probably of equal or greater importance than any other characteristic, the pest management specialist must exhibit a clear-cut, unbiased approach to the program. He should owe allegiance to no one except the grower for whom he is working. The free choice to utilize all available management tools—cultural, chemical, and biological controls, resistant varieties, etc.—for the long-range production of optimum yields and quality of all crops in the system should be his objective. Both he and the grower should understand this and be in complete agreement.

Literature Cited

MUELLER, A. J. and V. M. STERN.
1974. Timing of pesticide treatments on safflower to prevent lygus from dispersing to cotton. J. Econ. Entomol. 67: 77-80.
RAKICKAS, R. J. and T. F. WATSON.
1974. Population trends of *Lygus* spp. and selected predators in strip-cut alfalfa. Environ. entomol. 3: 781-4.
STERN, V. M., R. VAN DEN BOSCH, and T. F. LEIGH.
1964. Strip cutting of alfalfa for lygus bugs control. California Agriculture. 18: 5-6.
WATSON, T. F., L. MOORE and G. W. WARE.
1976. Practical Insect Pest Management. W. H. Freeman and Company. 196 pp.

Chapter 12.

CURRENT TRENDS AND FUTURE OUTLOOK

M. T. AliNiazee, K. S. Hagen and S. C. Hoyt

Broadening the Base of Pest Control

For many centuries man co-existed with the arthropods and plant pathogens that lived in his crops. Natural control, inadvertent selection of pest-resistant plant varieties, and the use of empirically discovered cultural practices allowed man to harvest enough food without greatly disrupting the ecosystem.

As planting of native crops expanded, and alien crops were introduced into new habitats, pest outbreaks became more frequent. Early attempts to suppress these outbreaks involved the use of a few heavy metal or botanical compounds applied directly on the pests or their feeding sites. The tactic of classical biological control was developed near the turn of this century and provided a means of coping with introduced insect pests. At about the same time deliberate selection and propagation of pest-resistant plant varieties were started. From the late 1800's until about 1940 applied entomologists focused their research on the biology and ecology of insect pests, and cultural practices were developed to reduce pest outbreaks. But in the early 1940's, when the first powerful synthetic organic insecticides came on the scene, most applied entomologists paid less attention to arthropod biology or ecology. Insect populations were reduced to exceedingly low levels by the new chemicals. Preventive pesticide treatments, based largely on the calendar, became the main strategy of pest control. The non-target fauna was scarcely considered, and pest populations were rarely monitored.

The general usage of broad spectrum persistent insecticides became a powerful selective force for any trait in arthropod populations that would allow survival. Not only did resistant strains of former pests begin to predominate, but arthropods which had been of minor significance in the crops became prominent pests because of ecological disruptions. Ecological relationships became apparent. The role of natural enemies in agroecosystems was particularly illuminated. It was necessary for applied entomologists to once again consider the biology and ecology of total crop fauna and to monitor pest populations.

Recognition of problems associated with past pest control programs has focused attention on the need for a change to broaden the base of pest control. Understanding ecological relationships in an agroecosystem has emerged as the primary goal of many applied entomologists, and this approach has provided us with more reliable control strategies. Such techniques, however, require a greater understanding of pest biology and ecology, which in turn requires much more research and supervision.

Future Outlook

Strengthening the foundation of pest management

The foundation of pest management must be strengthened through improved sampling and monitoring techniques for pests and their more important natural enemies. Improved methods of predicting population trends and the establishment of economic injury levels are essential. These techniques should then be used as the bases for making decisions on preferred tactics for suppressing pest populations and/

or their damage. In the past few years improved sampling methods have been developed for several important pests, but additional attention must be focused on those species where sampling methods are inadequate or impractical besides improving existing methods. Without good sampling methods and sound information on population dynamics, it is impossible to predict population trends accurately. Without the ability to predict population trends, it is impossible to determine whether or not economic loss is likely to occur even if information on economic injury levels is available. However, since these two elements are closely interrelated, the information on one will also be helpful in determining the other. Lack of adequate information in any one area will increase the likelihood of calendar spraying and decrease the application of pest management concepts. This does not imply that a pest management approach cannot be attempted before all information is available. Pest management is dynamic in that a program can start with a minimum of information and change through time as more information relating to the agroecosystem or new techniques become available.

Pest-suppression tactics

All pest-suppression tactics used during the past few decades need to be reevaluated for their suitability in pest management programs. The tactic of selecting plants for antibiosis against pests has been particularly successful in controlling plant pathogens, nematodes, and some insects. Any serious pest which is new or rarely under natural control, and which receives regular pesticide treatments, is a key pest and should be considered a top candidate for plant resistance research. Often by eliminating key pests, secondary pests resulting from the use of pesticides will come under natural control.

Natural enemies, including parasites (parasitoids), predators and pathogens, if present in a favorable environment, play important roles in determining the level of their host populations. Natural enemies present in the ecosystem must be preserved. If natural enemies are not present, or are too few to control pests, there are long-and short-term biological control tactics that should be tried.

Long-range programs center on classical biological control methods. If a pest is introduced it should be considered as a candidate for the introduction of its natural enemies. Furthermore, in pest management programs where exotic natural enemies have been introduced but not established it may be necessary to repeat the introductions, keeping biotypes from different climatic zones separate when culturing them for release. Also, genetic strains of natural enemies that indicate some degree of pesticide resistance should be introduced into agroecosystems where natural enemies more sensitive to pesticides prevail.

Several short-term tactics in biological control hold much promise, including culturing natural enemies. Insect pathogens, like *Bacillus thuringiensis*, which are produced on non-insect artificial media have become widely used as selective insecticides. This appears to be a prelude of what is to come when the baculoviruses of arthropods become registered and available for field use. The nuclear polyhedrosis and granulosis viruses, because of their specificity, should be among the most used pesticides of the future to control arthropod pests, particularly Lepidoptera.

Periodic release of predators and parasites can provide an effective means of controlling a wide variety of pests. This tactic hinges on being able to culture the predators and parasites directly on artificial media. Such colonization now is limited to a few species including *Chrysopa carnea, Trichogramma* spp. and certain phytoseiid mites, used mostly in greenhouses.

Increased use of natural enemies through environmental manipulation will receive more attention in the future. This can preserve or increase the effectiveness of predators and parasitoids. The use of behavioral chemicals to attract and change searching behavior of natural enemies is another tactic of the future.

In the past, sterile insect releases have been aimed at eradicating target pests. However. insects cultured for release may not be as competitive as individuals in the wild populations. Natural selection can reduce the impact of autocidal techniques. Also, if there is any genetic propensity in the target species populations to avoid carriers of lethality, such characteristics are sure to be manifested. In fact, selection for parthenogenetic strains may even result when sustained non-density dependent mortality pressure is applied to pest populations. The approach in the future will probably be to use the sterile-insect release method to suppress target pests rather than to eradicate them (unless the target pest is very localized and isolated). Other ingenious theoretical genetic strategies have also been proposed to suppress pest organisms. Deleterious genes can be introduced into pest populations to alter the habits of target pests so they will not survive in certain habitats.

The cultural control practices that have been employed in the past to reduce pest problems will continue into the future, but with more consideration given to intercrop associations. In this way it will be possible to maintain crop diversity and encourage establishment of natural enemies, particularly general predators. Although strip harvesting, interplanting, purposeful planting of non-crop plants and selective weed control have been shown to reduce pest problems, they have not been widely adopted. However these stabilizing methods are available for future pest management and hopefully their use will be expanded. Various cultural practices aimed at direct suppression of target pests, such as planting dates, soil preparation methods and harvest dates, should be further investigated and more closely integrated into management programs.

Chemical pesticides are presently the most effective methods of quickly suppressing a threatening pest population. In the future, pesticides will certainly be an important part of our arsenal for managing pests. However, application of a pesticide should be looked upon in the total pest management context, often as a last resort. Utmost consideration should be given to selectivity, dosages and timings so as to cause the least amount of disruption to the agroecosystem. Even though there may not be enough pesticides available with a narrow selectivity, the application of broad-spectrum pesticides, combined with specific pest attractants, will sometimes accomplish selectivity. Various economic factors such as the availability of raw materials, energy requirements, and cost of production and use of pesticides will have a great impact on future pesticide use.

With further development of insect-growth regulators, including juvenile hormones, perhaps these compounds can be used with selectivity. However, their use may be limited because of high cost. The isolation and identification of behavioral chemicals is in its infancy. As more pheromones, kairomones and allomones are developed, a wide array of new attractants and repellents might become available for managing and monitoring pests and their natural enemies.

Management decisions

Integrated pest management will become more complex because of its ecological base. The decision-making as to what strategies and tactics to employ to control pests could become the most important element of the whole program. Careful sampling and monitoring of a pest and its natural enemies will be a basic part of all decisions. Decision-making would have to be accomplished by a careful assessment of the whole situation including the pest potential, natural enemies, weather condition, harvest dates and marketing practices. Economic injury levels are now available for a few insect pests, but for most insects such information is inadequate or unavailable. Some decision-making can be accomplished at the farm level, but because of the complexity of agroecosystems most decisions will probably be made either by the county extension agents or pest management consultants. Some of these consultants now monitor only pest populations; others also monitor natural enemies, and some advise farmers on cultural practices as well as pest control. The advice they offer to the farmer is based on surveillance, available knowledge and experience. Each year this becomes more complex. In some localities Extension Service monitors weather conditions and warns farmers over large areas to be on guard against pest outbreaks. Such warnings in the past have been based mainly on past experience where, under certain weather conditions, certain pests often attain outbreak proportions. In more advanced examples, field consultants monitoring pest populations provide extension computers with pest stage and abundance data which are coupled with weather data inputs and, by using physiological day-degree summation techniques, attempt to forecast pest outbreaks (fig. 20). A recommendation to treat a certain crop with pesticides then involves many farms in the same climatic zone. Pests of any crop may, however, vary in abundance from field to field and from area to area. Therefore, each individual field should be monitored and the decision to treat one field, or parts of a field, should be based on the advice received from pest management consultants.

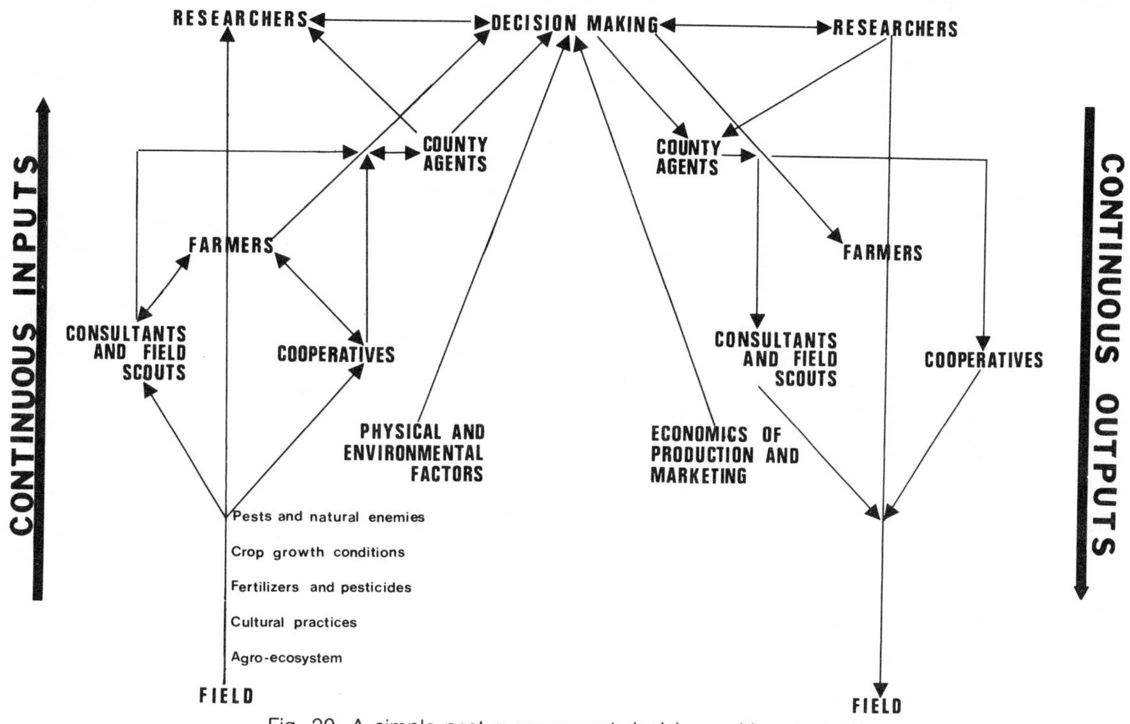

Fig. 20. A simple pest management decision-making model.

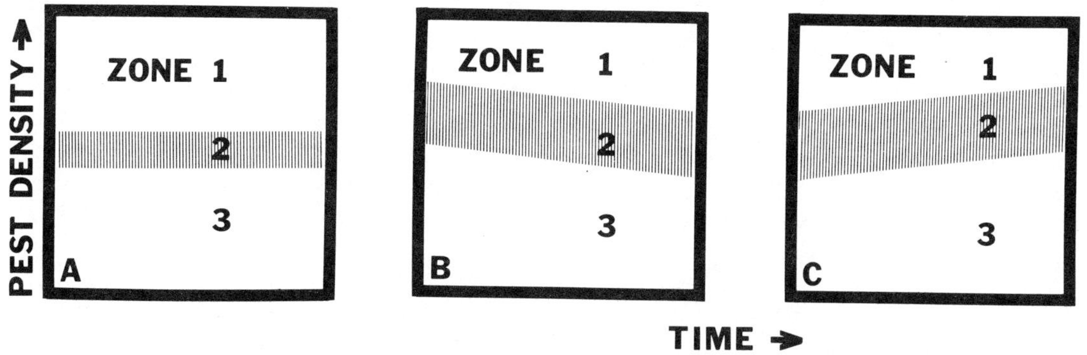

Fig. 21. Zone treatment situations (see text).

If the present trend continues, there will be forecasts of pest outbreaks emanating from centralized University Extension Service headquarters based on inputs to computers which will aid in pest management, but the most reliable advice for decision-making will come from pest management consultants monitoring individual fields and pest situations. Such decision-making will be relatively easy for distinct "treatment" and "no treatment" zone situations where pest populations are either too high and economic loss is certain or too low to cause any substantial damage (fig. 21). When populations are in the risk zone, decision making becomes complicated, and extensive deliberation and continuous monitoring (sometimes daily) may become essential. With some insects it may be possible to tolerate a relatively high number of individuals early in the season (fig. 21, B), while in other situations it may be possible to tolerate more insects later in the season (fig. 21, C). The risk zone itself may fluctuate, depending on natural enemies, weather conditions, and other environmental factors.

Research in pest management

A great deal of research is currently being conducted on tactics to suppress pest populations or reduce damage and most likely this will be continued. It is unrealistic to expect other tactics to rapidly replace pesticides as the principal method of controlling pests of many crops. A greater incorporation of selective pesticides, or dosages which are compatible with other tactics, can lead to an improved environment for further pest management advances.

Recent developments in insect physiology, behavior, and communication have opened up new directions for pest control research. Sex pheromones and juvenile hormones are two of the many promising tools which will probably receive increased emphasis in the future. Although sex pheromones of a large number of major crop pests have been identified and artifically synthesized, much basic work on behavior and field application of these pheromones needs to be done. Major emphasis should be placed on understanding the complex behavioral response of field insects to pheromone traps, including analysis of the effects of weather. Attempts should be made to develop economic-injury information based on pheromone-trap catches.

Research on cultural control will get more emphasis in years to come mostly because of the relatively small cost involved in such control. As the energy crisis deepens, availability of pesticides will decrease and cost of production and use will be much higher, thus giving incentive to investigation of various cultural methods of pest management. An integrated approach combining all these methods will no doubt have a much greater appeal to future researchers and farmers.

Education

Education in pest management concepts and practices is a continuous process. As new information becomes available and is put into practice, many established procedures may have to be changed. Therefore a dynamic but simple and practical approach should be preferred for teaching pest management.

As information required for an integrated management program becomes available, it must be transmitted to those who will apply it under field conditions. It seems unlikely that a management program with a moderate degree of complexity can be operated by most farmers. To provide the specialized training required to implement these programs, several universities and colleges are offering degrees or training in pest management.

If maximum use of integrated pest management is to occur, it will be necessary to mount a major effort to educate farmers. Unless current pest control practices are failing or become too costly, farmers will be reluctant to change from established and accepted methods of pest control. A well-organized educational program must be developed to provide information on the need for and benefits of the pest management concepts. It is only when farmers are convinced of these values that the trained pest management consultants will be hired. The number of people trained as pest managers is small at present, and likely will increase only according to the numbers of jobs available.

Another educational avenue which could be explored is that of altering consumer attitudes toward minor blemishes on produce which affect appearance only. Acceptance of these blemishes would relieve the pressure toward controlling all pests, even those causing minor damage, and would allow much greater latitude in the application of pest management principles. However, no organized attempt is being made to change these attitudes. On the contrary, the competitiveness of today's markets seems to make consumers more reluctant to accept these minor injuries. Governmental agencies often set the quality standards which force more pesticide use.

Legal restriction on insect parts in processed foods is another deterrent to maximum use of pest management. Fragments of beneficial insects are as serious as those of pests. In many instances, farmers are forced to substitute pesticide residues for small quantities of insect parts in order to eliminate any possible "contaminants" and sell their crops. On many crops, both the consumer attitudes and legal restriction on insect parts demand an intensive schedule of pesticides for compliance. Moderation of attitudes and restrictions would allow the substitution of more ecologically sound methods of regulating pest damage.

A great deal remains to be accomplished in research and education on pest management, but a start has been made as the information presented in this publication indicates.

Benefits of integrated pest management

Integrated pest management can lead to less expensive production of food and fiber, a substantial reduction in pesticide residues, a greater stability in vegetation systems, and an improved environment.

Greater stability in crops, parks and gardens is expected if integrated pest management is practiced because full advantage is taken of natural controls operating on all organisms. Natural control is encouraged by the long-range strategies of introducing missing natural enemies, developing pest-resistant plant varieties, and utilizing cultural practices that encourage diversity. These strategies reduce reliance on pesticides. When it is necessary to use a pesticide, the least disruptive chemical is selected to preserve as much stability between organisms as possible.

Environmental pollution by pesticides is reduced as a consequence of integrated control, and energy and chemicals are conserved. Under todays integrated control approach, the role of pesticides is being reversed from that of yesteryear. Pesticides today are considered within a total context and often become the last resort and not, as in the past, the first tactic. Because the essence of integrated pest management is applied ecology, the long-range economies of all organisms, including man, benefit.

References

Clark, L. R., P. W. Geier, R. D. Hughes, and R. F. Morris.
1967. The ecology of insect populations. London: Methuen & Co. Ltd. 232 pp.
DeBach, P. (Editor)
1964. Biological control of insect pests and weeds. London: Chapman and Hall, Ltd. 844 pp.
Hall, D. C., R. B. Norgaard, and P. K. True.
1975. The performance of independent pest management consultants. Calif. Agric. 29: 12-14.
Huffaker, C. B. (Editor)
1971. Biological control. New York, London: Plenum Press. 511 pp.
Jones, D. Price, and M. E. Solomon (Editors)
1974. Biology in pest and disease control. Oxford: Blackwell Scientific Publ. 398 pp.
Maxwell, F. G., and F. A. Harris (Editors)
1974. Proceedings of the summer institute on biological control of plant insects and diseases. Jackson, Mississippi: Univ. Press. 647 pp.
Metcalf, R. L., and W. H. Luckman (Editors)
1975. Introduction to pest management. New York: John Wiley & Sons. 587 pp.
National Academy of Sciences
1969. Insect-pest management and control. Washington, D. C.: Publ. 1695. 508 pp.
1969. Pest control strategies for the future. Washington, D.C. 376 pp.
Pimentel D. (Editor)
1975. Insects, science and society. New York: Academic Press. 284 pp.
Rabb, R. O., and F. E. Guthrie (Editors)
1970. Concepts of pest management. Raleigh: North Carolina State Univ. 242 pp.
Stark, R. W., and A. R. Gittens (Editors)
1973. Pest management for the 21st century. Moscow, Idaho: Idaho Res. Round., Inc. 102 pp.
Stern, V. M., R. F. Smith, R. van den Bosch, and K. S. Hagen.
1959. The integration of chemical and biological control of the spotted alfalfa aphid. Hilgardia 29: 81-101.
Summers, M., R. Engler, L. A. Falcon, and P. Vail (Editors)
1975. Baculoviruses for insect pest control: safety considerations. Washington, D.C.: Amer. Soc. Microbiol. 186 pp.
Whitten, M. J., and G. C. Foster
1975. Genetical methods of pest control. Annu. Rev. Entomol. 20: 461-76.

1.5m-rep-4/93-WJC/ALS